The
Billionaire's Chef

Cooking for the Rich and Famished

The
Billionaire's
Chef

Cooking for the Rich and Famished

Neal Sheldon Salisbury

CRC Press
Taylor & Francis Group
Boca Raton   London   New York

CRC Press is an imprint of the
Taylor & Francis Group, an **informa** business

To Fred Lozier, for nurturing my interest.

To my parents, Charles and Brenda Salisbury, and also to friends Maria Karlsson and Gary Kowalewski. I could not have had this career or written this book without all of your support.

To Bob Edmondson and Dickey Unangst, for showing me how much fun it is to be the chef!

FIGURE 1   NEAL SPEAKING (PHOTO CREDIT = SUSAN OLIVER). UNLESS OTHERWISE STATED, ALL PHOTOS ARE FROM CHEF NEAL SALISBURY'S SCRAPBOOK.

CONTENTS

A DAY IN THE LIFE

"Just push on through," I say emotionlessly from the passenger seat of a small SUV as we race up to the bumpers of two cars traveling side by side down Cable Beach in the Bahamas. I have $1,500 in cash and a Visa card in my pocket. It is the year 2001, and Nassau is still limited on gourmet food supplies. Back in the kitchen the meats are thawing and the crème brûlée is on cooling racks. I have nothing else.

I am a little sick to my stomach as we roll into downtown Nassau. We storm a major grocery, which can be totally hit or miss on the very basics; but French butter and decent olive oil will be there and either great or horrible produce, nothing in between. I am in and out in 20 minutes. Next stop will be Greycliff, the island paradise's only 5-star restaurant. As we pull around back, the polite British estate managers, a couple in their 40s (I am 32-years-old at this point) say, "I do not think you can go in the back like this; maybe we should see the maître d'." Without responding, I am in the back door in a flash. I deal with the sous chef in the head chef's absence and pull out all $1,500 and fan it in the air: "I need demi-glace, white truffle oil, chicken stock, Gruyere cheese—I don't care how much it costs!" I treat him like he is a king because right now he is. We settle on a fair price, and I slip him an extra $100 for his troubles. Next stop: Potter's Cay, to have a giant grouper killed in front of me. Four and a half hours till plates need to hit the table.

Four hours earlier, I step off the plane from Ft. Lauderdale with my tool-kit and a few suitcases filled mostly with music equipment and casual clothing. A nice Bahamian man puts my bags in the car. Sun shining and hot, with windows down we casually make our way to Lyford Cay, a tax shelter for impossibly rich expatriates from all over the globe and a super safe community complete with armed guards, 13-foot stone walls, and its own police station. Once inside, there is no real reason to go outside. They have their own school, post office, stores, and private banks. It is a beach peninsula with a mix of castles, walled-in manor houses, tasteful bungalows, and unique pleasure palaces surrounding a sprawling golf course all anchored by a large beach club that seems to be the hub of all things social there. Billionaire Peter Nygard and Elle McPherson live on one end in a massive Polynesian-style home with a Mayan temple waterpark on their private beach. Neighbors include Sean Connery in a modest bungalow on the golf course and some European royalty. Five of the residents are on the Forbes world's billionaire list, and another five hide just enough assets to stay off it. We chat up the guard as he checks us in. Driving past the private marina, we get a view of yachts up to 200 feet standing at the ready for the owners to show up by golf cart. The private airport (Million Air) is only 20 minutes from the gate where Gulfstream and Challenger jets sit with red covers over their engines to keep the sand from blowing in.

We pull up to a British colonial cottage, and it's hard to tell the size as most of the house is hiding behind tropical landscape. It is very Palm Beach and tasteful. It is one-third the size of what I am used to working in, but it oozes charm. I am introduced to the housekeeper, Ingrid, a kind-looking Bahamian woman in her mid-40s, and am taken upstairs to a gorgeous suite with cathedral ceilings, a big fan, California king-sized bed, and marble bathroom. I am told Ingrid will do my laundry and make my bed. I like this already. We go downstairs for a tour of the house: six bedrooms, a pool, and we are right on the golf course. We get to the kitchen. "It's a little cozy, but I will make do," I say. He replies, "What are you talking about? This is *your* house; the estate managers will pick you up in half an hour."

Six weeks earlier and shortly after arriving home from a season on a yacht in the Mediterranean, I got a call from my estate agent in Beverly Hills. I live in Ft. Lauderdale, Florida, where I play in the sun between my jobs. My gigs typically run from two weeks to three months each. My agent says a family in the Bahamas is looking for a really good chef and asks if I would I like to be submitted for it. "Sure," I say, but I really want to go back on a yacht. For weeks I get brief calls from personal assistants representing the family from their offices in the United

States: more references, background checks. Will you take a drug test? Every so often it's something new. In the yachting business I am used to taking the call and being on a plane within 48 hours. This is different, and I have no idea about the gig. They don't give any details about the family, just in and around the location, of which there are a lot of different situations. One day I get a call, and the person asks, "Can you do a trial for a week?" and "Can you cook dinner for 10 people the next day?" I say cooking isn't a problem, but I won't have time to shop. They tell me to fax them a shopping list and the estate managers will procure everything I need. The meat is already there from New York, so I say, "Fine," and get packing.

The estate managers pick up me and my tools in a golf cart. I am a confident man and a very good cook; now I am one with sweet digs. We turn the corner, and there is this brand-new hotel. I ask jokingly, "What is that?" They say, "That's the house, 11 weeks old." I had seen a lot in my few years as a private chef and living in South Florida, but this was literally built as a palace (as confirmed later when I saw the blueprints titled as such). The main house is one level: three bedrooms, 31,000 square feet, and over 300 feet long. Five other buildings behind the walls include a 12-car garage and four massive gates to let you into different parts of the property. The service wing and dining room are 6,000 square feet. It takes over a dozen paces to cross the kitchen. I am nervous for the first time in my career. I had a few billionaires under my belt by this time, but these were Russians and there was nothing humble about them. I am introduced to their daughter, who is 5'11", rail thin, and looks like a Russian Paris Hilton. She is gorgeous and very nice, maybe 25 at the time. She will be my liaison with the family about all things food. The estate managers are very nice and easy to get along with. They are a practical couple that oversees the building of the house and writes out the checks. It is the single most expensive house between North and South America. It looks a lot like the Atlantis resort at the other end of the island and shares the same contractors. "Mr." comes into the kitchen. He is in tailored slacks and golf shirt sporting lots of gold jewelry and smoking a cigar. In a heavy accent, he comes up to me and says, "Tonight is very important to us, so we appreciate you coming so quickly." He hands me the cash and the Visa and leaves the kitchen. I am introduced to the butler and under-butler, two immaculately groomed Bahamian gentlemen about my age in white dinner jackets and tuxedos. They are polite and conservative.

So far, everything is good except that I know the stakes are a lot higher than anything I could have gotten my head around. This will prove to be a theme throughout my career as a chef to billionaires.

The estate managers show me around the servant's areas, the locker room, the laundry: bustling with interior staff, all Bahamian and wearing white uniforms that remind me of hospital attire. The walk-in refrigerators are all empty except the freezer, which is teeming with Lobel's prime, dry-aged beef. This lets me know that I had better bring my A game.

Dinner begins promptly at 7:00 pm. It is now a little after 1:00, and they ask if I have had lunch yet. They would like to take me to a friend's café. We head back upstairs to the kitchen, and I ask if there is anything they could not get on my list. They look at me and say, "What list?" I am a nearly 6'2" New Englander pushing 250 pounds. When I get serious, I don't have to say much for things to get real—and things just got real.

Being a private live-in chef at the highest level is not an easy thing to become. It takes years of schooling, great cooking chops, great people skills, career savvy, and you have to live cleanly to pass all the drug tests and background checks. You have to become a professional traveler. Eventually you will have to change languages, currencies, and types of transportation, sometimes several times a day while on the road—and most of all in the beginning you need luck.

It is like Hollywood: you get yourself an agent, get your paperwork in order, go to interviews and industry events, and pound the pavement trying to get a break. You stay in crew houses with like-situation people and either help or try not to kill one another depending on how much you remind each other of yourselves. I used to get up at the crack of dawn, ride a rickety crew house bike five miles over I-95 to a string of dirty shipyards, then sneak past the security guy who only sometimes turned a blind eye or bought my absolutely unbelievable story of what I was there for—to beg for day work, which is scraping hulls and painting bilges so I could get a day's pay and lunch and could prove my work ethic to maybe get hooked up with a real job in my department. In only two years I had made it in the private chef world. I had a rep as one of the best chefs in the industry and was in a palace with one of the richest families in the world, my own house, car, maid, and fat cash. All of my agent's eyes would be watching how this gig would go, as I'm only as good as my last one. Now, every second counts, as the clock ticks, and I have no food yet!

When you travel to another country for a job like this, if it does not go well (i.e., you fail) even if you were doomed from the start, you have to stay there until it is time to fly you out, which can be very uncomfortable. You lose your income, your home and perks, the friends you made there, and the regional identity you were building. I did not want that.

I make crème brûlée and pull the meats from the freezer while the estate managers get the car. The second the brûlées come out of the oven, I jump in the car: "Just drive! Seconds count, and I have the money if we get a ticket." One thing to always remember in the private chef world is that there are no excuses; no one cares why dinner is 20 minutes late. My ex-girlfriend, who is also a yacht chef, used to say "G.T.F.O.T.T." (Get the Food on the Table). I would add, "Even if a few villagers have to die along the way."

Upon arriving back in the kitchen from our adventures in town, I promptly get to work. It is a massive undertaking. I picked a very ambitious menu, and it is all served butler style at this level. Butler style is when the butler offers you a silver platter with utensils on it, you take as much as you want, and he does not serve you. When the butler serves you off the platter, it is called Russian service. Royalty feels that Russian service is considered a form of portion control and that it forces you to eat the recommended portion for everyone, not the individual. The Queen of England eats butler style. Remember, they eat butler service at almost every meal, and there is always a second offering. My fish are served boneless, skinless, and as whole sides. Meats are done sliced in bite size or two, if fork tender. No gristle, no bones, unless serving a proper chop.

On tonight's menu:

Chicken consumé with a quenelle

Kasa auflauf (9" Swiss Gruyere cheese soufflés)

Field green salad with warm port wine vinaigrette, Granny Smith apples, shaved carrot, toasted walnuts, and Carles Roquefort cheese (aged in French caves)

Chateaubriand paired with a sherry-shallot, creamed demi-glace, truffled duchess potatoes, and haricot verts

Sides of jumbo Bahamian grouper with a citrus-mango buerre blanc, basmati rice, and sautéed spinach

Crème brûlée

Cheese board

FIGURE 2   PRIME CHATEAUBRIAND.

FIGURE 3   CHEESE PLATTER.

While I am testing equipment, trimming and reducing, I make small talk with the butler, who looks depressed at best. He tells me that there have been 11 chefs in 11 weeks: "some last one day; some a week." The daughter breezes in and tells me that the cheese soufflé is all they are talking about as "Mrs." is from Switzerland and she has a few Swiss guests. No pressure.

The butler informs me that the house next to mine is Sean Connery's. I ask if he ever comes over for dinner. The butler says that the family think's he is a good actor but not of their ilk, so no.

I finally have three minutes to run to the locker room and change. Stress and heat have given me diarrhea. Sometimes it is only in a toilet stall that you can get your head out of the game to slow your roll before you run yourself off the cliff. I don't have much time left, and there is still so much to do. Everyone around me is silent as they pass through the kitchen. I have been given a scullery maid from Sri Lanka to help with pots and pans during dinner service. She, too, was of the mind-set that I would not be there for long. Nobody made eye contact throughout service. You just heard a horrible buzzer go off anytime "Mrs." wanted the butler, which was every five minutes. Proper dining tables have a doorbell mounted under the head of the table so that the primary can summon staff without anyone knowing. It is silent in the dining room but sounds like a state execution in the staff areas as a grating, electric shock honk that makes hair stand up on the back of your neck. It is this way in every estate and on every yacht. You never get used to it, but it does give you a sense of urgency.

The butler lets me know how courses are flowing and when it is time to platter the next. I lightly garnish platters and crocks and place them into a two-way, glass-door warming oven that opens in the kitchen and on the other side in the pantry, where they can be removed and taken to the dining room.

Course by course, the food makes its way out as I warm and finish the next. I am numb and unhappy, physically drained to the core and surrounded by obvious faces of doom from the staff. This must not be going as well I thought. When the kitchen is almost finished being cleaned, the butler comes in, looks me in the eye, and says, "The princess liked your crème brûlée."

I go back to my house, pour a stiff drink, curl up, and watch TV till late in the morning alone on the floor in a giant guest bedroom.

The following morning, to my surprise, the family is waiting for me in the kitchen when I arrive. "Mr." hands me $5,000 for shopping and says that dinner was impressive and that they felt "proud." To date it was the finest executed meal of my career—unbeknownst to me all with a princess, a prince, and three billionaires at the table. If you are going to make it to this level, you must be able to rise to the occasion when it counts.

If you think this book is going to be all fun recipes and polite-smile inspiring stories, you are wrong. The life my peers and I lead is filled with challenges, disappointments, and sometimes unexpected triumphs. We are constantly pulled from our comfort zones. Our safety and sanity are pushed to their limits. The battles of ego, the criminals we sometimes work for and live with (every James Bond villain needs to eat), the money, and the sex (or sometimes lack of it), coupled with loneliness, all come with the job. We are artists with a benefactor, not sellers of food, and expectations are stratospheric. Pockets filled with cash, we are the decision-makers. When it's good, it's very good. When it's bad, it is very bad. We are given the resources, and there is no one else to blame.

I hope you will find the following chapters an honest and informative journey deep inside the ultraprivate world of being the chef to some of the world's 2,000 richest people. I have been the chef to over 20 of them in my 15 years of private service. I have rarely felt comfortable with this career choice that started at sixteen with a willingness to grow in my career and a lot of common sense. I have taken this journey a long way from growing up outside of Boston. I once said in earnest to a colleague, "I don't think I was meant to be a yacht chef, but I definitely think I was meant to have these experiences." Now I would like to share them with you. Enjoy.

PREFACE

My name is Neal Salisbury. I would like to take you on a journey into a world that really exists but few have seen or even heard about. For over 15 years, I've spent my days and made my living in this world. I will do my best to give you an honest, albeit surreal, view of what it is like to have worked as a top private chef during the Golden Age (1998–2008) of professionally crewed private luxury yacht building/cruising and grand estate renovation/building.

There will be stories, recipes, tips on event planning, and social etiquette. You will get career advice from someone who has been to the top of the game, made mistakes, and groveled his way back into the game. If I tell a story, I try not to use a name; if I use a name, I try not to tell too much of the story. This is a very private world, and I will not betray the trust of the people who have brought me into their homes and private lives. Their names are not important to the story. There certainly will be juicy stories to let you know what really happens behind the hedgerows and on those big white boats that line the marinas of Monaco or St. Barths—you just simply don't need to know the names.

Who am I, you ask?

I am a classically trained chef who began at age 16. I grew up in Nashua, New Hampshire, about 40 miles outside of Boston. I had a nice upbringing. Dad owns a small research and development company that also manufactures a line of high-tech electronic measuring devices. Mom is a homemaker. They have been married for over 50 years, and I'm the middle child of three, with an older brother and a younger sister. Nice people.

We had a great culinary program in my high school, whose instructors were Fred Lozier, a former bakeshop and patisserie 1 & 2 instructor from the Culinary Institute of America in Hyde Park, New York; and Carmine Lovergine, a retired sous chef from the Waldorf Astoria. There was a carpeted, candlelight dining room open to the public. No one had that back in the 1980s. We were so lucky. These instructors greatly influenced my career path.

New Hampshire is not a culinary mecca by any stretch. People live quite well there, but they are Yankees first and place a high value on big portions and a rustic ambiance over complex flavor profiles and exotic ingredients. I was 14 when I first realized I might like to work in a restaurant one day. My mother took my siblings and me to a steakhouse in downtown Nashua. The dining rooms had dim amber lighting and smooth jazz in the background, and the smell of well-seared steak permeated the place. We had a pretty waitress sporting form-fitting black trousers and a thin white blouse that allowed the amber light to bring out her skin tone and the outline of her white brassiere. I remember thinking that it all seemed like theater to me and that the experience aroused all my senses. We had certainly been to nicer places before, but that night a seed was planted in my head.

So, at age 15 I started my career at McDonald's—not as a cook but at the drive-through. To this day I have never seen such efficiency in everything but the actual employees. I still admire how upper management can make a well-oiled machine run so well using a group of people with no real restaurant training. I learned there that battles are won in the planning stages of a restaurant, not by flying by the seat of your pants. Not enough restaurant owners realize this. I then moved over to a fine dining restaurant up the street from McDonald's to be a dishwasher after school during my sophomore year of high school. It was so much fun there because four of my best friends and I all got jobs at the same time. The kitchen was big, they fed us well, and the waitresses were pretty. Restaurant culture is fun; it is like a family and is hypersocial. I fit in well with the fast pace, and we could blast the radio in our little corner of the kitchen.

After that summer I moved to cooking on the line at another place up the same street. I spent two hours in culinary class at school and cooked

FIGURE 4   CHEF NEAL AT THE CIA IN 1987.

at night four or five shifts a week. I played in a band throughout high school and was popular with the girls at a local church youth group, so no one should feel bad for my childhood.

As luck would have it, we had two chefs from Switzerland relocate to New Hampshire to open six good restaurants that gave the locals access to real food from scratch and, for those of us lucky enough to work for them, some European techniques and sensibilities. This was critical in our early development as chefs.

I wanted to go to music school in California in the worst way. I was a good musician, but my no-nonsense Yankee father would have none of that. So he made me an offer I could not refuse—a Datsun 240Z sports car with all the trimmings and a promise to pay the lion's share of my choice of culinary school. So off I went to New York State and the prestigious Culinary Institute of America (CIA) at Hyde Park. Fred Lozier's letter of recommendation and my father's checkbook sealed my fate.

I had never been away from home and my friends for any length of time. People in New Hampshire are generally nice and polite. In New York, it was like being thrown to the wolves. The CIA in the 1980s was not fun. It was and, after several trips nearly around the world now, still is the most competitive environment I have ever experienced. It was a seven-hour

FIGURE 5   NEAL AT 16 JUST STARTING HIS CULINARY CAREER.

class day plus five hours a night of homework including eye-numbing Beta videos on old techniques. I experienced rude tristate-area people who trashed their own dorm bathrooms. Female enrollment was 23% of the student population, three-fourths of whom shaved their heads to a buzz cut for practical reasons. To this day I never give any hassle to a girl who graduated from the CIA. If they could handle it there, they are tough!

But I loved the food and grandeur there. Six-course lunches with Russian service, wine with meals, and celebrities eating in our restaurants—all in a former Jesuit seminary. I worked washing pots in the American Bounty and Escoffier Restaurants at the school from the beginning. I ate up the information and was always above average in the technical and the scientific subjects, both of which are valued there.

The big difference between Europe and America in becoming a classically trained chef is that in Europe you start your training at age 13. You become a stagier, a neighborhood kid who stands outside the kitchen door of the local bistro, and if the chef likes you he will invite you in to wash dishes and give you supper in return. Once you prove your work ethic, the chef will move you on to peeling veggies and then prep cook with a stipend and meals. Eventually he may write you a letter of recommendation to get yourself an apprenticeship with an experienced and willing chef for school credit. Eventually you will become a chef de partie, running a station on the line. After a few years, if you show responsibility and leadership you will become a sous chef, a sort of vice president of a kitchen, your last supervised stop before becoming a head chef. The sous chef does a lot of hands-on cooking, makes the schedule, and is in charge during the head chef's absence.

In Europe, the governments laid out the cities 500 years ago to be practical and to last. A corner of a town square would have been designed to be a restaurant for the working man, and today it still is. Chef jobs do not turn over much, and there is a long line of people waiting for them. Both chefs and their patrons have a more serious attitude about their food. Most chefs in Europe have trained under only a few chefs but for longer and with more accountability.

In the United States, restaurants are cash cows and rarely last longer than 20 years. Concepts and trends change quickly, and most turn over chefs every few years as they burn out trying to please an insatiable, disconnected audience at all times of the day and night. The odds of the chef at a restaurant caring enough to invest his valuable time trying to teach you anything other than his own menu and personal philosophy are not in your favor. Training begins at 16 in a high school vocational program and working in restaurants after school to gain hands-on experience to practice the craft you learned during the day in a safe, controlled environment. Your instructor's whole focus is the transfer of knowledge. It is a very consistent and efficient way to advance your career. You miss three years of paying your dues to your local community, but other than that we American chefs misses nothing in the way of learning during our formative years over a European. It is simply just a cultural difference.

The main reason people go to school is for food science and discipline; there is no guarantee you will learn those on the streets. The CIA was the best in the world for both. I took the information in like a sponge, but at 19 I wasn't ready to be in that environment. I left at the end of the first year right before externship. I had already done the CIA bakeshop and pastry studies in high school. I would go on to study wine in Bordeaux, France, and had heard the Escoffier and American Bounty lectures many times while scrubbing their pots.

I do not regret this as I was luckier than most and had a lot of options ahead of me.

In the years after the CIA, I owned a high-end ice cream and dessert shop in New Hampshire and spent a ski season as a waiter at the prestigious Balsams Grand Hotel and Ski resort on the New Hampshire–Canadian border. I moved to Hartford, Connecticut, to go into the music business and, funny enough, went to music school for three months before getting bored with it. Eventually, I wrote for two music magazines and owned a 10-piece Steely Dan tribute band. I then moved to Boston to put on concerts for the record company that owned one of the magazines I wrote for, and I brokered high-end musical equipment between professionals.

During this time in New England, I always worked as a waiter or bartender a few nights a week at each area's best restaurants. I socialized with the chefs and got all their recipes along the way. I love this time in my life because I made so many lifelong friends. When you are a musician, you do not judge other musicians on material status even public morality; you judge them on talent and loyalty. Foodservice workers could really learn from this.

It was during this time that I took up sailing on the Charles River. I had a flexible schedule with the concert promotions and working a few nights a week at the restaurants. I sailed almost every day and took every possible class. I had a knack for it and eventually volunteered as an instructor. Sailing has played a big role in my life, as you will see later.

FIGURE 6   NEAL WITH ELECTRIC GUITAR.

One day I got tired of the whole music business, because once you reach your regional potential it is very hard to take your career national. It just hit me one day that I had been very lucky and that maybe I should close that chapter in my life and put my culinary education to work. I had the CIA alumni want ads sent to me and looked at becoming a sous chef at a nice inn in rural Connecticut. The chef said that he had just filled the position but that he had a friend in the Caribbean with a few restaurants who was looking for someone good; a week later I was living in St. Croix in the U.S. Virgin Islands.

I ended up working at several different places during the year I spent there while bringing my cooking chops back up to speed. I loved the Caribbean—joined a traveling sailing race team, fell in love, swam every day, and sat in with bands. In the Islands, it's all fun in the sun during the day, and dinner is taken very seriously. People dress for it. I really enjoyed working with people from all over the world, not just from the northeast United States. They brought new ideas and ingredients to the table and with the exception of the only other

FIGURE 7   DRIVING A SCHOONER IN KEY WEST, FLORIDA.

CIA guy on the island (we are now friends), they were not combative and competitive but fun and supportive. We danced around the line and enjoyed the time spent over the stove. And then, when we were done with dinner, the nights were filled with dancing and gorgeous sex.

I worked hard to become the best possible asset to a sailing race team and was fortunate enough to be considered for our local America's Cup challenge at the time, so I moved 40 miles north to St. Thomas. Soon after arriving, the team expanded and moved to a different island, leaving most of us behind. So then I moved five miles over to the beautiful paradise of St. John. I worked for a restaurant group that owned most of the best eating establishments on the island. I was rented a room in a famous house at the top of Bordeaux Mountain that was owned by a tour manager for U2 and Paul McCartney. There were gold records on the walls and African art everywhere, and the house was 3,500 square feet with five balconies all up on mahogany telephone poles hanging over the edge of the cliff some 1,200 feet up overlooking the Sir Francis Drake Passage. Four of us rented the house together, and I think we paid $2,600 a month in 1998. I had a dirt bike and a Jeep CJ7, and I leased a horse from my friend who owned a pony camp. My girlfriend was young and beautiful; I was on the island's top sailing race team and taught children's sailing on Saturday mornings. I had known real quality of life at this point. Times are different; you could not live that way there now on a cook's salary.

One Sunday night, after a grueling weekend of sailboat racing at the annual Rolex regatta in St. Thomas, I came back to work. My eyes were like a reverse raccoon's from four days of racing, and I was exhausted. The bartender went downstairs to fetch a bottle of wine, so I delivered a plate of food, which I never did, to some customers at the bar. I was a recognizable fellow back then as I was tall, slim, and tan with long hair in a ponytail. All at once a girl lit up and said, "Hey, you're that guy from the Mirage."

I said, "Yup."

She continued, "You guys nailed that start. You ducked and covered and hit the wind line perfectly right out the gate! Did you make this?" as she looked at the plate I was presenting.

"Yup," I replied.

Handing me her card, she said, "Hi, I'm Denise, and I am a crew agent in Newport, Rhode Island. Have you ever considered mixing the boat thing with cooking?"

"Nope," I said.

She then added, "Think about it and call me if you are ever in Newport. I can make that happen."

A few months later the season slowed down, and since I had not been home in two years I went back to see my folks in New Hampshire. A week later I got on a bus to Newport. This was the last time I would ever be a normal person or have a normal life. Thank God!

When I arrived in Newport I was a really good cook and a competent mariner. I was tan and pretty and a bit untamed from having spent so much time in the Caribbean. I met with two agencies in town who crewed up the boats (often back then a one-woman show with a secondhand desk, a phone, and a fax machine who worked from a seasonally rented storefront). I was told immediately by the first that if I did not cut my hair and get a decent resume together they would not even speak to me. Denise was happy to see me but basically said the same. I thought from my sailing days that this is what a sailor should look like. Walking the docks in town, I realized that I was in a different world: one filled with big white motor yachts. They lined the docks everywhere I looked. I realized that I was not going to be working on a sailboat and that I better listen. By nightfall my shoulder-length, sun-drenched locks were gone. I found someone to help with the resume and taped a drugstore passport photo to it. I put it through a color copier, and the next day with my ego slightly more in check was sent on foot to four job interviews, none of which I got.

Two weeks later, one of the agents called and said that she wanted to see if we were doing the right thing. She sent me to Connecticut to join a beautiful 84-foot Burger yacht, a finely crafted boat with a really nice galley, similar to a kitchen in a fancy home. This boat was worth about $4 million and for that time was a good-sized yacht. It was Labor Day weekend 1998. The owner had gone from making a few hundred thousand dollars a year working for himself to selling his company for over $100 million. I spent the weekend with him and his family in the Hamptons and Fisher Island, New York. A short charter took over the vessel, on which 12 guys spent the whole trip on cell phones and ended with a celebration dinner at which they announced that while onboard they had purchased a major basketball franchise. This was so exciting to me. I was driving fast tenders, cooking and eating well, and we were partying with the owner's son. These were epic nights. At the end of eight days I was handed cab fare, a bus ticket to Rhode Island, and $1,766 cash. This included my day rate and a 20% gratuity from the very satisfied client. I had never seen this kind of cash for a week's work and play—which I would have done for free.

With a good reference, the agent agreed that I would be successful in yachting if I stayed focused. After this I was off to Fort Lauderdale, the world capital of yachting, to enter the big leagues.

Now you know how I got started and that I was lucky. I had all the tools to succeed but had to leave my world behind and adapt to this new one. The industry was happy to get a nice, presentable, real chef who had never done drugs and was not an alcoholic. These jobs were, however, not easily obtained.

In 1998, almost all professionally crewed yachts (about 1,700 of them) were about 100 feet in length. The few billionaires out there may have had a 130-footer, and the rogue sultan or two may have had a monster-sized converted ship that sat mostly in the Mediterranean but did not move much. In those days, the boats were harbor hoppers and went out mostly for long weekends and the owners would enjoy local restaurants along the way. Most of the chefs were local girls whom the captains were dating as the owners just needed a cook who could pull off a gourmet meal once in a while. However, some owners wanted consistently excellent food onboard and hired trained chefs.

When I began my career, each of these private yachts featured a crew of four or five with a captain, mate, chef, and a stewardess or two. Captains and mates had weekdays to do their own engineering. We had fun. Chefs earned $4,000 a month ($250 a day) for freelance, all in cash, and we had crew cars and room and board. When the boats chartered, we got a 10–20% gratuity based on the value of the base rate of the charter. Good money and good times for fun in the sun. Boats chartered for between $18,000 and 60,000 a week in those days, and tips were always in cash.

This is exactly what I signed up for. I wanted nothing more than good food, a cool job, travel, and an awesome social scene. And it was all about to change drastically.

What was going on in the real world was that information technologies were coming of age. Cell phones, personal computers, and the Internet created entire new industries that created totally new jobs and ways to mass market. The stock market was accessible to the common man. You could not lose: the wave of IPOs and major companies were gobbling up small companies and ideas at such a rapid pace that you ended up with a mountain of overnight millionaires, and those who were already wealthy were elevated to the stratosphere. I do not know how much the millennium factored in, but it felt like a lot. At this time too, people were so rich so fast that they could not hide it, and it was, for the first time in our history, okay not only to talk about but also to flaunt! I believe this is because absolutely anyone had the opportunity to advance at the time by just buying stocks or a spare house. Many captains at the time became millionaires this way, especially with the conversations they were privy to while traveling with their employers.

There have been economic booms in the past and certainly a few expansive eras in yacht building, but for some reason—and I don't know anyone who knows why—the ultra wealthy and the very rich started building yachts. They ordered them at a rate that no one could keep up with. There were simply not enough skilled, experienced yacht builders in the world to satisfy the demand. Within the first few years of my yachting career, the median boat being built was 177 feet and required a crew of 14. When a yacht owner received his new build, he sold his old one at a profit and often gave the captain a cut for helping with the sale. He would also take the crew off three 100-foot boats, as they were the only people in the world with any yachting experience. There became a huge demand for crew, and in our industry time is the commodity. What does the owner have a lot of? Money. What does the owner not have a lot of? Time.

There was no time for crew member training or backgrounds, which led to a lot of problems during the growth of this industry. Agents now had panel-lined offices in new low-rise buildings with secretaries and banks of computers for crew to check in online and make resume updates on the spot. If you were a good chef you could name your price—and we did. If you worked freelance, a trained chef was pulling $400 a day, and boats were now chartering from $100,000 to $500,000 a week with the average trip lasting two weeks. This amounted to serious tips. Here we were: no house or car payment, no training in money management, afraid to use a financial advisor for fear that our international, tax-free cash would lead to our demise.

We were living rock star lives. Excess was a badge of honor: 22-year-old stewardesses buying multiple bottles of Dom Perignon for the table in a St. Martin nightclub and pulling cash out of their bra to pay for it. Tips were split equally between the crew members. We were paid vastly different salaries for our responsibilities, but anyone could make or break a charter. If I get a $10,000 tip, so do you. This creates valid pressure among the crew, as we are all shareholders in the tip.

I think it fair to say that most of my peers and I have also worked in the owners' estates from time to time as well as providing their aircraft with victuals, and we will exhaust this subject later.

A superyacht or gigayacht is the single most expensive object you can spend money on for pleasure. No one needs a yacht. There is nothing you can do on a yacht that is more efficient than doing it somewhere else, except to create the most amazing experiences and memories possible to man.

A private jet can take you to five locations in a day at record speed with no delays to stop a crisis to make sure that 50,000 people don't lose their job next year. When the jet is not in use, it sits in a hangar with covers on the motors and will last 30 years with routine maintenance. Most maxed-out private jets cost around $30 million. You can get a used light jet for under $1 million, and a tricked-out 747 will cost you several hundred million. I recently photographed a $1 billion yacht in Germany that is one-third complete. It will require a massive crew every day that lives onboard forever. Being over 600 feet and dockage costing mostly over $10 a foot nicely included in a 10% of total value annual operating budget—you can imagine the expectations of the chef who gets that job. I certainly don't want that.

Between 1998 and 2008 the number of professionally crewed private motor yachts over 100 feet in the world grew from about 1,700 to just under 5,000. And they were much bigger. The industry quintupled in size during the first 10 years of my career. It was so fast paced and so unregulated that most of us spent the better part of a decade running on adrenalin.

We department heads (mate, engineer, chef, and chief stewardess) were the point-of-purchase decision-makers. For any of us to have the boat's American Express Black Card and $5,000 to $10,000 in cash in our possession was not unusual. Vendors wined and dined us. I have been to parties that cost $500,000 with Swiss watches given away as door prizes.

I learned early on that I was a very good chef but that I had a big personality and was a little hard to live with or deal with for any length of time, so I chose to work freelance. It paid more money and was a tougher job, but I was in and out in a few weeks or months and was so busy that I stayed out of trouble for the most part. Even crew who did not like me knew that if I was in the galley on charter that the tips most likely would be high. I also did not have to sit around during yard times and boring stretches behind an owner's house. I eventually marketed myself as specializing in large or difficult programs as I liked to solve problems that were not my own and was very good at logistics.

FIGURE 8   YACHTS AT PARADISE ISLAND, NASSAU, BAHAMAS.

I had my own life off the boat, and when the phone rang the hair stood up on the back of my neck. When they called, it was not to chat; it was go time. I kept a storage unit near the airport, and when it was time, toys in, tools and uniforms out, ready for my patdown and takeoff. I only got 2–48 hours' notice to fly anywhere in the world. I became a professional traveler, and of the hundreds of jets I have been on I was almost always by myself.

It can be lonely: 15% of the time, it is like being the James Bond of cooking—your mission, should you choose to accept it, is to fly commercial to Europe, change to a private jet, get picked up in a Mercedes by a man who barely speaks English, and get whisked off to a chateau where you will be creating the best food of your life for someone who is trying to impress a fellow captain of industry. For the week, you kill it and make love to his personal assistant the night before you fly out, dropping her phone number in the trash as you board your return flight in Zurich; the other 85% of the time it is like being the Dirty Harry of cooking, where some evil or unorganized program has a firestorm going on and they need you to be firefighter, at least in the catering department.

Sometimes you were with backstabbers, drunkards, and the fringes of society that have only a fellow drunk relative to thank for a job that they in no way, under any circumstances, should have gotten.

You learn to watch your back and not become a scapegoat, and gorgeous sex and fun of any kind are off the table. You just get the job done and collect your pay. These situations take time to get over and to come down from. Often afterward I checked into a motel before coming home just to sleep for three days and let the adrenalin subside. After that it feels good when the wire transfer comes through and your friends take you down the beach for Sunday-Funday.

THE RICH AND
THE WEALTHY

Comedian Chris Rock once explained the difference between rich and wealthy: "Shaq is rich. The man who signs Shaq's paycheck is wealthy." He is referring, of course, to NBA legend Shaquille O'Neil. I don't believe I could explain it any better than that.

There are somewhere around 450,000 families in the United States with a net worth of at least $10 million. Any of them can charter a nice yacht or rent an incredible villa in an exotic location. These are the people who live in those great oceanfront mansions and stay in the world's top hotels. There is no meal they can't afford. They have cool car collections, and every once in a while they charter a private jet to take them on a getaway. Trophy wives are common, they are on the boards of every local charity, and their children look like angels. Other than a special occasion or an off-season charter, my peers and I do *not* work for the vast majority of them.

After 15 years in private service including the decade of the proliferation of private service (1998–2008), I have in some combination lived with, dined with, cooked for, traveled with, or put on private events with well over 20 of the world's billionaires, all while collecting a paycheck. My clientele also includes A-list celebrities such as Martha Stewart, Diddy, and Jerry Seinfeld. I am not famous. Some of these people never knew my name or if I was the regular chef or not. Billionaires are not calling yacht charter companies saying, "I have got to have Chef Neal on my trip!" And thus far the Bravo network has not offered me a hilarious over-the-top show about my travels and hijinks. I was simply considered the right guy for the job when it was important, when the regular chef was unavailable, or when it was an out-of-control situation and the program was hemorrhaging chefs.

When people find out what I do for a living, they inevitably ask me one of two questions: "What's your best dish?" or "What are they really like?" referring to my high-profile clientele. Are they happy? Are they mean? Are they really good-looking? As to the first question, we will cover my culinary preferences in the following chapters. As far as billionaires go, I have found them to be male, female, pretty, ugly, nice, not nice, fun, boring, pretentious, unassuming, happy and unhappy. Those are human issues, not money issues, and if you are not one of them then human issues and putting your pants on one leg at a time are about all you have in common with them. Someone once asked me what they all have in common with each other. My reply? They all have to eat.

I hear the word *billionaire* almost every day on TV. It has become part of our vernacular. Here are some numbers to try to get an idea of how few billionaires there actually are.

The estimated global population is roughly 7 billion people. *Forbes Magazine* says that in 2014 Earth was home to 1,426 billionaires. These are the ones that they can track, mostly from stock holdings and traceable assets. *Forbes* would be the first to tell you that if a billionaire does not want to be on the list, all it takes is a quick call to their legal and accounting departments and you would be hard-pressed to find them in print anywhere. They also do not count colossal family trusts, as they look at the trust, from a legal standpoint, as an investment business and its trustees as a group of less wealthy people. Those who track wealth estimate the actual number of the world's billionaires to be closer to 2,000. No one will ever know for sure. So the world's billionaire population stands at about 0.0000002% of the population. That's not one in a million. That is one in 5.7 million. The good news, the odds of becoming one of them are still better than winning a sole Powerball jackpot. So at least you have that going for you. At the *Forbes* 1,426 number, they have a combined net worth of $5.4 trillion, and the average list member has around $5 billion—but who really knows?

Who are they? They are sultans, oil sheiks, Third World leaders, royals, captains of industry, CEOs and their children, venture capitalists, hedge fund managers, and people who married up and are no longer married. As a whole, the Walton family (owners of Wal-Mart) often tops the list at a combined net worth often between $100 billion and $120 billion. They are surprisingly much less conspicuous than many of the folks who share a spot on the Forbes list. All billionaires in history made their money the same way—control and expansion—even the ones who acquired their fortune through marriage.

Let the number $3.7 billion sink in for a moment. Next time you are in a major city, look up at the skyscrapers around you and try to figure out how many of them that would buy. All that steel and glass towering 500 feet or more into the heavens with the marble foyers and brass elevators—then throw in all the surrounding hotels with all their luxury appointments. It would take you a while. This is the net worth of these people: the cash-out-and-run-away number. Corporate assets for some are closer to $1 trillion. That's a lot to be in control of: massive power. So when they come home or take off on their gigayacht to some remote island, they expect a lot from those of us lucky enough to travel with them, and they should. Even those who are making their money through crime and deceit still work hard at it and are paying good money for our services. If their lifestyle turns your stomach, you can always find a mutually convenient time to quit.

You do not have to be a nice person to make or inherit a monster fortune. I have worked for arms dealers, master corporate thieves, massively immoral types, and a few that would make your skin crawl just by looking at them. All of their checks cleared.

I do not judge myself by the way these people live or by the way they treat me. Their lot in life and mine are not linked. I am simply there to provide a service that my peers and I are uniquely qualified for. By that I mean not just the cooking but also fitting in with the owners and their guests. We are occasionally requested to socialize with them. This is done to promote a better rapport between the staff and guests and sometimes simply to fill up the mock nightclub we set up onboard. We sometimes dine, go scuba diving, or genuinely enjoy a good Karaoke night with them in the main salon of the yacht. Luckily, the vast majority of people for whom I have worked were lovely and just trying to enjoy the fruits of their labor.

When I first came into the business, I looked the part; I was a cool enough customer and a seasoned diner. I did not, however, speak the language or really understand the magnitude of who I was at the table with. I am not wealthy. I was not brought up in society, and I brought my middle-class sensibilities with me. We were all God's children and created equal. Right? This is something that can be frustrating to someone who has paid his or her dues to become a major domo, a yacht captain, a department head, and certainly a financial tycoon.

Spending time out socially with a green staff member who does not know their place and does not show proper respect can be an exercise in patience and wasted time that you will never get back. When the average person comes into this world, they do not have enough letters in their alphabet to describe these new experiences. They struggle to adapt and find their comfort zone. So many become almost hostile to authority figures and start telling them how to do their job in an effort to get others to adapt to them instead of the other way around. They need to distance their personal beliefs and learn to relax and be polite. We are not looking for unique individuals during working hours; that's why we wear a uniform. The art of being a good private worker is to make the client feel at home and to let them speak as freely as they desire without judgment. There is no billionaire on Earth who knows my political beliefs.

We always have to be careful when staying out late with them and remember that we are on the payroll. Never outdrink them. Your favorite color is whatever theirs is. When you are on the payroll, you are not really their friend—even if you have a relaxed relationship or get along great with them. The number-one rule in private service is, "Never forget who pays the bills." If they invite you to visit them after your employment with them is complete, they really like you, and you are then a friend. This is a nice perk and doesn't happen every day. Enjoy it.

Billionaires have serious security concerns on all fronts. They are high-value targets in the lucrative kidnapping trade. They and their family members all carry kidnap insurance. The bad guys know this. When a high net worth person is apprehended, they are usually detained unharmed, a reasonable ransom is demanded, and the insurance pays out. The victim, not wanting negative publicity, does not let this hit the papers, and you will never hear about it. Their trauma and recovery are lived out behind closed doors. Sounds unbelievable? What if you own a newspaper and your daughter is kidnapped at 19, brainwashed, and joins a domestic terrorist gang who make her rob banks while caught on security cameras with a rifle in her hands? Sound familiar? Or what if your oil baron trust fund son is kidnapped at 16 and you receive one of his ears in the mail along with a ransom note? Believe me, the Hursts and the Gettys are far from the only billionaire families that have had misfortune fall on them at the hands of others. John Lennon was another stratospherically wealthy fellow who had another type of security risk lead to his demise right outside his front door at the hand of an obsessed madman.

Billionaire Edmond Safra was killed accidentally at home in Monaco in 1999. His live-in male nurse, in a bizarre plot, was trying to look like a hero by setting a small fire in a trash can outside Mr. Safra's bedroom. The nurse had the intent of coming to his rescue to gain favor. The fire got out of control. Safra and a staff member were cooked alive in their own living quarters. I was working for a friend of his at the time. We were in the Caribbean and monitored the situation constantly.

I have worked for more than one person with a standing hit put out on their lives. These are just some examples of the security threats that these families face every day, even from within their homes. Doris Duke's death is believed to have been hurried along with the aid of her gin-blossomed, barefoot butler, who subsequently inherited most of her fortune. She was the richest woman in the world at the time. He started wearing her clothes after her passing.

That's a lot of public grief for just 1,226 people. You will never hear about their private tragedies.

I bring this up for a reason. Before you demonize this group for conspicuous consumption and find them tasteless for not handing that private jet over to the poor, consider that these people run the world. Their decisions make the economy move. They employ tens and even hundreds of thousands of people. Even with that jet, someone built it, maintains it, pilots it, fuels it, insures it, governs its regulations, and cleans it.

A lot of these people cannot just fly commercial or stay in a normal hotel or go out alone in most areas for security reasons. They surround themselves with a beautiful cocoon of comfort, more or less gilded cages for some to be protected from the dangers of the outside world. There are those who just don't want to face other's reality. I am not making excuses for them. In truth, I would much rather be rich and somewhat anonymous than brand-name wealthy.

As far as the trust fund babies are concerned, babies are not angels who pick their parents. I know many from the "lucky gene pool." No one asks to be born into a poor family, and no one asks to be born into a rich family. Birthright can be cruel in both directions. You just play the hand you are dealt.

But enough drama. Here is a list of what the average billionaire on that list owns:

- A primary residence in a major city, unless retired; then it would be moved to an international luxury tax shelter (figure $30~200 million but the most expensive house in the world was recently completed in India at over $1 billion).
- A country house ($12 million), a very large compound or farm where the extended family can be together for weddings and special occasions ($25–80 million)
- A beach house ($14 million)
- A chalet or supercondo in a ski resort area ($12 million)
- A chateau or castle in Europe ($35–200 million)
- A yacht of no less than 164 feet (the famous 50 meter cutoff to be a superyacht at $50–100 million) or a gigayacht of up to 600 feet (yup, it is being built in the Lürssen shipyard in Germany at $1 billion)
- A dozen reasonable houses for the kids and casual guests ($22 million)
- Many tasteful smaller boats ($7 million each)
- Several helicopters ($5 million each)
- A light jet ($12 million)
- A long range jet starting at a Gulfstream G-4 and topping out with a private Boeing 777 with two stories and a dance floor ($30–300 million)

FIGURE 9   AUSTRIA CHATEAU.

FIGURE 10   NEW LÜRSSEN GIGAYACHT BEING FINISHED IN GERMANY.

FIGURE 11   ROMAN ABRAMOVICH'S PRIVATE 767 DWARFING OTHER PRIVATE JETS.

FIGURE 12   ROMAN ABRAMOVICH'S 377-FOOT GIGAYACHT PELORUS. THE COLOR MATCHES
HIS PRIVATE BOEING 767 WIDE-BODY JET.

- A few have submarines in the 60 feet range ($80 million)
- Paul Allen, Richard Branson, and Elon Musk own spaceships (not kidding, and no clue on price)
- Of course, you are nobody in this exclusive club without owning a major sports franchise ($200 million to $2 billion)

They hide money in jewelry and art, so I will move those to the investment column and will not cover cars since they don't really add up to much comparatively unless you are Jay Leno or Prince Rainier.

So you're starting to get the picture. But there is good news! Billionaires are not the only ones who live this way! Look at this list of possessions, and be sure that the vast majority of people on this list will have most of the items on it and super consumers like Paul Allen and Roman Abramowitz (both tied for the title of "King of Yachts") will have all this and more. They are modern-day Vanderbilts.

I always say, "The closer to the fire, the more likely you are to get burned." Somebody has to make all this stuff work, manage it, maintain it, move it around, and clean it. If you like expensive stuff and lots of action, you may like working for the wealthy. We live in their homes, drive their cars, fly on their planes, go on vacation with them on their yachts, eat their food, drink their wine, and when we travel with them we get most of the same privileges that they get. It is at times absolutely over-the-top amazing. The perks can be unbelievable. Everything is amplified at this level. The highs are very high, and the lows are very low.

The hallmark of the wealthy is that they are, as a group, extremely mobile. If you work for one of these people, you will at some point have to travel. Whether you work in an estate and the owner moves all of the upper staff to their Palm Beach compound 10 weeks a year or you travel with the owner wherever they go or perhaps you work on their yacht and are always on the move, you will definitely need a passport.

FIGURE 13   NEAL WITH THE BOSS'S LAMBORGHINI.

The biggest jobs are usually live-in positions for convenience and security reasons. This is kind of cool because your housing, food, car, and gas are all supplied, and you will live in beautiful surroundings, all with no commute. The hard part is that you lose some of your freedom and your privacy. You, like them, are victims of security. There are card keys for everything, high walls, security checkpoints, armed guards, and huge gates, and you will almost always have cameras on you when you are outside. It does not usually feel like a military zone, but it is all there for your protection. It does not, however, allow for a lot of wiggle room after a random wild night out.

You also cannot bring people over to your place without prior consent, and they must be really important to you, like a spouse or immediate family member. Remember, you are background checked, referenced, and drug tested. Your cool new friend from the wine bar is not. You live in their home, not your own. All this protection that you enjoy makes it very tough to have a normal love life. I have lived at marinas, estates, and behind locked gates with polite, armed guards for most of my adult life. I am very safe. I am also single. I would not date a girl who is constantly on call to leave on a few hours' notice for up to several months at a time and you could not visit her.

We also get used to living in that world where we constantly enjoy experiences and privileges far beyond our pay grade. When we go home, our friends are plenty glad to see us but do not seem to care about our tales of how we got to deliver the boss's Lamborghini to Key West as much as we do. And if you have a slow season, it surely will be brought up by someone at your weakest moment. One captain said that on trips back home he used to recount the details of his trips, and people started to think of him as a braggart. He said eventually he learned to talk only broadly about the itinerary and bring up only popular, well-visited tourist attractions—not the good stuff. We genuinely want to share with our friends the good fortune bestowed upon us; however, it just brings distance between you and your normal friends. They are never going to really be able to get their head around what we really do. It's like telling them you were abducted by aliens. With less than 5,000 megayachts, superyachts, and gigayachts on the planet, the odds of someone having seen, toured, or vacationed on one of them is only slightly better than them having any of those experiences aboard a UFO—although there have been a lot more documentaries on TV about UFOs than superyachts. People know about UFOs; you can talk about them all day with anybody. Finding a mate who understands what you do or feels comfortable socializing and networking in this field just puts more obstacles in the search for an appropriate mate. Still, somehow I can't imagine my life without having had these experiences.

At the end of the day, the way these billionaires live is interesting. I have never been bored in my career. The fine craftsmanship of everything they build and buy, the availability of beauty everywhere you look, the global adventure: they celebrate life with the best food and the finest wine. They throw unforgettable parties, and their pioneering spirit is inspirational. It is the human aspect of these programs that makes it awesome. Otherwise, it's just a collection of objects. It is what you do with them. They are just a means to an end. It's the memories that we make with them that makes it special.

Please enjoy this book. It is being written with passion to take you on a journey. Whether you use it for escapism, as a practical career reference, or to help you throw better events doesn't really matter to me: just enjoy the trip.

CHAPTER 1

The Routine

Congratulations! You have scored your first job as a private chef to the wealthy. You have beaten the odds and are about to journey into a world few have seen. You are excited but a little scared too; after all, most people are a little afraid of what they don't know.

The first time I ever crossed the gates into an estate valued at over $20 million, it seemed like a whole other planet with its own atmosphere and gravitational pull. It was in Palm Beach back in the 1990s. I felt like I had to ask permission to do everything, even though I was actually going to be left alone to look after the property for the next two days while the local staff took their last few days off before the season began. It was so spectacular: priceless works of art, Roman artifacts, life-size Greek statues, and rooms that were assembled in France then broken down and shipped to the home for installation. I roamed the property the first night barefoot on acres of putting green grass down past the Koi ponds to the docks of the Intracoastal Waterway and watched the boats go by. I wondered what this adventure was actually going to be like.

I spent the first two days taking inventory, cleaning, and receiving orders before the upper staff were scheduled to arrived from Manhattan for the next 10 weeks. About a dozen of us would live in a wing in the house set up like a dorm. We would each have a small bedroom with a TV and a sink. Common bathrooms with water closets and showers were situated down the hall; the whole setup was much like a one- or two-star European hotel. Back in Manhattan we would all be issued proper apartments for the rest of the year. The estate managers had their own apartment on the property, and many of the maids and groundskeepers were local townspeople that commuted and did not have housing provided. We ran the property with eighteen full-time staff and hired out extra helpers when throwing large parties.

On the third day after my arrival, the upper staff came on a private jet with the boss and his wife. She was an old-school head of household straight out of the gilded age—not only a royal chaser but also an actual friend to royals, who frequently visited! They had a special Rolls Royce called a Carmague that was just for visiting royals; it was kept in the British garage off the main house, which was around the corner from their German garage. Mrs. was a pleasant but demanding older lady who was known as a consummate party thrower. Her main goal in life was keeping her husband happy. They had been together since they were young and poor, and she idolized him.

My mornings began with a shower, turning the oven on, having a cup of coffee at my breakfast nook in the main kitchen and a phone call to three fish markets to see what had come in before dawn. This gave me final options for the day's menu, which already had meats, side dishes and pastries factored in. I then made fresh muffins and croissants and baked the breads that had been rising overnight. Mr. and Mrs. put a breakfast order in through the butler, which is when she would review the day's menu. After breakfast, she came to the kitchen, and we went over the choices and any possible guests and finalized a game plan. At this point I headed out to pick up the fish and anything else I needed. Meats were kept in a freezer and were shipped in from New York. We had accounts with most of the markets, and I had a few hundred dollars in cash and a credit card from the boss.

Cruising around Palm Beach in a nice staff car picking up fresh foods from nice people in nice shops with the sun on your back is not a bad gig. My understaff of a sous chef and a scullery maid ensured that life would not be too difficult on a daily basis. The sous chef made a good but simple staff lunch, which was eaten at a long table in a common room where staff took breaks and did odd projects requiring a big table during the in between hours. I prepared lunch for the family and guests, and my sous chef and I worked on knocking out a dinner prep list throughout the day: dessert projects, stocks and sauces, etc. In the afternoon, the kitchen staff got some downtime to take care of personal business, watch a movie, or just take a nap. At 6:00 p.m. (most staff ate at noon and 6:00 p.m.), I put up a buffet staff meal (salad, starch, meat, and vegetable) and then got onto the serious business of getting our mise en place (French for things in place) or meal at hand prepped while we put out hors d'oeuvres for cocktail hour. Dinner was mostly three courses served by the butler and his understaff: a starter; an entrée of meat, starch, vegetable, and a sauce on the side; completed by a dessert. We cleaned the counters and stove, and the scullery maid took care of the pots and floors. I worked on a menu for the next day, which took into account the day of the week, previous meals, the weather, and possible company. After this it was off to the shower, relax for a bit, and maybe catch a drink at one of the many great bars and restaurants the island had to offer—where the bartender knows more about you than you do, keeps your poison on file in their mind, and beats you to your seat with it. Early to bed, early to rise, and repeat for the next 10 weeks.

Palm Beach is much like many other wealthy seasonal resort location where the family is more at play and not commuting to an office on a regular basis. Santa Barbara, California, and Newport, Rhode Island, have similar schedules. No two gigs are alike. It is, after all, private employment. There is no union, organization, or federal agency governing someone's home or boat. There are different levels of expectations from employer to employer. Some want a good cook who can pull off a gourmet meal on occasion, and some want a chef with a Michelin star tied to their name, straight out of one of the world's best restaurants. Some will want breakfast, lunch, and dinner; some a mix of meals; and some entertain constantly at all levels.

I had another job where the boss worked all week in Manhattan and retreated to a weekend house in the Hamptons. There are great food cities all over the world: all food is sold by the pound, and the best foodstuffs usually go to the highest bidder—tuna and Kobe beef to

Tokyo and everything else to New York City. It is kind of unique in the way it operates for private chefs. If you are not familiar with the area, it can be very intimidating. Once you get clued in, though, it is awesome. The absolute best of everything is available on a tiny, 11 × 2 mile island. The trick and fun are in finding your preferences. The standard of quality in NYC is very high, and the better neighborhoods have excellent local markets.

For this job, the butler made a set breakfast for the boss, and I arrived later in the morning, having purchased freshly made breads from a top bakery and fresh fish I picked up on the way in. I would start a dessert project and prep for lunch. Meals were carefully crafted for the family, with the boss coming home from the office for lunch a few days a week. In New York, the staff took care of their own meals for the most part and lived out. During the afternoon I hit the markets for incidentals a few days a week, filling a cart with dry goods and dairy and just putting in a list with the butcher and deli. I then left it and walked out the door, took some personal time, and headed back to the house. You see, in Manhattan it is not practical to keep a taxi outside while you shop or try to drive yourself. Everything is delivered. We set up accounts with the stores, and once we have our goods pulled the store tallies, bags, bills, and delivers. There is a delivery charge, but it is cheaper than operating a car and finding parking in the city. The parking rules are complex, and they send an army of wreckers around 24/7 with a license to tow at will. The impound on the west side highway looks like something out of a *Dirty Harry* movie, and it is not a place you want to visit twice. I know this because I have been there twice. Your goods will arrive at your door in a reasonable amount of time, and if the order is not a big one some of the other staff may even put it away for you, provided you regularly spoil them. If it is more than three bags or boxes, head home and take care of it yourself. You don't want to be "that guy." Then we prepared for hors d'oeuvres and dinner; three courses was still the basic rule unless there was a special occasion. A full cleaning of the kitchen, and it was off to meet friends or head home.

On Fridays I headed out to the Hamptons after lunch. Most families keep a staff car in a garage in or just outside the city. I loaded up anything I needed from the city and hit the highway for the 100-mile trip, which can take anywhere from 90 minutes to seven hours depending

FIGURE 1.1 SAILING IN NEWPORT.

on your aggression level and traffic. The family drove out as well, with the boss arriving by chopper later in the evening. They usually ate out on Friday night, or if they really wanted to dine at home on Friday I left in the morning and they at out for lunch in the city.

The Hamptons are a collection of small farming villages at the eastern tip of Long Island. You know when you have arrived because the big four-lane split highway shrinks and goes from elevated to ground level. Everything slows its pace, and you peacefully roll through small towns that ooze charm and quality. Great restaurants, epic beaches, and world-class nightlife are everywhere. Endless designer shopping sprees and celebrity sightings are interrupted only by boat rides around Shelter Island. Most of what you need as a chef is found in the gourmet markets, butcher shops, and fish markets. There is a Williams-Sonoma and a few independent high-end culinary stores. The trick is to avoid Route 27 (Montauk Highway), which runs all the way through the Hamptons, if at all possible. Traffic does thin out east of Amagansett, but most of what you need is west of there. The heart of the Hamptons is from South Hampton to Amagansett, where there is a decent farmer's market and a really good fish market. This whole region is an intricate spider web of farm roads that are less densely populated than you might imagine. They are so totally fun to blow off steam on while putting your staff car through its paces (and I'm not talking old station wagons here—some of my staff cars have been a supercharged Range Rover, Jaguar XK8, and Mercedes SL convertible). You will definitely rack on a few extra miles doing this, but watching the corn vacuum in toward your after-draft in the rearview mirror beats sweating it out on the Montauk Highway with the tourists. There are back ways to all your destinations, and crossing Route 27 is not that tough.

Families like to dine in and entertain in a beautifully relaxed way in these retreat locations. Nantucket, Martha's Vineyard, and Cape Cod are similar in the way you operate. Working in Europe is a little different as the culture and their way of eating is not quite the same as in America. Typically, Europeans stay up later and rise later than Americans. You still head in early and do your baking, but then it's off to town to hit the markets early. The main difference is that in Europe we tend to plan the menu in the market itself. Farm to table is more the norm, and Europeans tend to grow and consume much of their food and wines regionally. Perhaps it is because the cost of fuel is high there and before the Euro currency exchanges and border crossings were obstacles for shipping. European towns have lived off local resources for centuries; only specialty items are shipped in. Because of this, the food is fresh, and this is a source of pride. In Europe, the early bird does get the worm, so we plan the menu and buy the ingredients on the spot before the townsfolk deplete the supplies. For the best service, show respect and use all the words you know of their language before humbly switching to your native tongue. This will really be appreciated and open many doors for you.

Regional produce items are plentiful, but if your menu gets ambitious you will need to have your homework done in advance and probably drive a little. A GPS in the car makes short work of getting there if you looked up the address first—especially if you do not speak the language. I was once based in a shipyard at Bremen, Germany, which is industrial and old-school. I frequently had to make trips to the city of Hamburg, which is hours away, to get the hipper products and services more common in a major city. Hamburgers also spoke a lot more English. Along your journey, you will find many culinary craftsmen who locally produce specialties and pastries that take years to perfect and require unique or well-seasoned equipment to produce. You may

FIGURE 1.2 FRUIT STAND IN BORDEAUX, FRANCE.

choose to add to some of them to your personally baked offerings to add to the authenticity of a regional menu.

Back in the kitchen, the butler will bring you a hot breakfast order and the day's menu to the boss for approval. It is now time to get ready for lunch. This is where you will notice the biggest difference from your work in America. Lunch may be served later than you are used to, and in most European countries it is the big meal of the day, especially in Latin countries like Spain, Italy, and France. They take the midday meal and subsequent nap to the extreme. Meals can be many simple offerings or classic courses, but you can almost always expect wine to be served daily at this time. The morning shop is not always the final shop, and this can be tough when most shop owners take that same meal and nap. Most businesses close for siesta: a two- to three-hour rest in the middle of the afternoon. It is very much a Mediterranean thing; you can't fight it, and you must learn to adjust. You may want to nap now as well, as you still get up early but the evening meal will be much later than in other parts of the world.

There will usually be a cocktail hour that will require hors d'oeuvres. Later a light, slow-paced, three-course dinner will ensue. Dinner will be different in every country and for each boss. You may choose to offer a pasta course, a fish course, and a small dessert in Italy or perhaps a light soup, a gigot of lamb, and profiteroles in France. European dinners should feature smaller, high-quality portions and simple but elegant presentation.

When dealing with towns near the Mediterranean Sea, unless you have a commercial account they are generally not set up for volume orders. If you are having a function, you must give plenty of notice to your purveyors. Most families buy what they need for the day fresh and do not keep freezers full of meat and fish. You will not find many big-box membership stores in most coastal areas. Small, high-quality shops usually look at me in horror when I ask for a kilo (2.2 pounds) of thinly sliced, individually hand-wrapped prosciutto or 10 kilograms of a meat or fish. When it comes to fish, the Mediterranean is a poor sea, having been overfished for thousands of years. You really need to get to the fish market early for anything good that is locally caught. The fishermen line their catch up under tents as soon as they dock, and you will have plenty of competition. No one at the fish market will care who your boss is. This can be frustrating; however, you can't beat the freshness of boat to table, and it makes you really appreciate your product.

It would be hard to beat a Caribbean vacation when faced with pure sunshine, endless rum, year-round boating, and crystal clear waters so perfect in temperature that when you dive in all you feel is wet. The local cuisines, the friendly smiles of the islanders so eager to cater to your every whim: what could be better? A Cruzan bartender explained the reality of the Caribbean to me in a simple joke:

> A man dies and meets St. Peter at the gates of Heaven. St. Peter tells the man that Heaven doesn't work the way he was taught growing up. You not only get to choose Heaven or Hell, but you even get to visit one first. The man says, "Show me Hell, that way I can see how bad it really is." In an instant the man appears in Hell. Beautiful women in tiny bikinis stroll along the beaches, people are dancing in the streets, and

well-dressed bartenders are serving the most decadent cocktails imaginable. The man appears back
at Heaven's gates, and St. Peter says, "What do you
think?" The man says, "I have seen enough; send me
to Hell—I want to live in Hell!" The man is sent back;
only now there is poverty beyond his understanding,
junkies are dead in the streets, and a hurricane is on
its way and visible in the distance. The man turns and
asks someone on the street what happened. They said,
"Before, you were just visiting. Now you live here."

There are many challenges to working in the Caribbean. It can be
tough to get the ordinary things in life done; you can't just fight your
way through without frustrating yourself and annoying the locals.
There is a natural flow to island time. If you are going to make it in the
islands, you must learn to pick your battles. In the Caribbean, other
than babies, everything is shipped in. Weather can delay ships delivering the vast majority of sundries. If the Heineken boat doesn't show on
time, the island's remaining Heineken supply becomes more valuable
than gold. Without good alliances, your boss may have to go without. That means they will probably replace you with someone who
knows better how to get hold of Heineken in tough times. However, the
weather is most often in your favor. You really only have two obstacles:
provisioning and traffic.

Traffic, you ask? There are three kinds of traffic jams in the islands.
Cruise ship meccas like St. Martin and St. Thomas can have 25,000
extra sightseers on the streets, both on foot and in every available taxi
and bus, when a ship is in port. This, combined with holiday rental
properties and an onslaught of yachts filling every marina slip and their
occupants grabbing every rental car and hitting the streets, can turn an
otherwise 25-minute trip to the other side of the island into hours of
tropical frustration. Next is the ever-present herd of goats blocking the
road. This is not unusual, and do not underestimate how long it takes
them to move along. Statistically, only 3% of goats care about your
car's horn. Try to nudge them with your bumper, and, after the ensu-
ing goat–car battle, you will never end up getting your security deposit
back on your brandless Chinese rental car. The goat jam is God's way

of telling you to just sit back and finish your cocktail. Finally, and most frustrating, is when two island natives who have not seen each other in hours happen to run into each other a few blocks from their homes. One is always in a crappy car and the other is walking—that is, until two quick beeps of the horn, which is a signal to come to the middle of the street, lean in the window, and catch up. When you have a tight schedule or are on the way to the airport to pick up the boss, this can make your blood boil, and you may feel the need to leave your vehicle and educate them on your plight. Don't. These conversations last a specific length of time, only known to islanders. If interrupted, I am almost positive that they start over. A big smile coupled with a polite, superlight double tap on the horn should remind them that you are waiting without the restart. Anything more would be social suicide.

The moral of the story is to plan ahead and leave plenty of extra time when traveling on an island.

Provisioning is somewhat unique in the Caribbean. These days, larger islands will have much of what you need in stock. The problem is that by the time the food arrives on island it has been handled a lot and has had a long, often turbulent boat ride. The food has been through customs, often with delays. If it looks good in the market, you can use it that day and perhaps the day after. If you are filling a yacht or are stocking up for a week at the villa, you will need to use a professional provisioner. These are local experts who will have your food flown in fresh and exactly to your specifications. A huge part of being a private chef is maintaining the owner's standard of cuisine anywhere in the world. This is not cheap, but it is the cost of doing business in paradise. This will be discussed in more depth later in the book.

You can, however, get almost anything you need in a pinch. I would load my pockets with cash and head into the kitchens or to the concierges of top resorts. If you use up your superyacht or monster villa cache with a humble attitude and play the needing-their-help card, you will usually come out on top. Do not go in with a huge list expecting them to do the job you should have done a week ago. They are usually happy to help and make a connection in the private service world. Always give them your personal business card, and a respectable tip goes a long way. The latter is not a percentage of what you spend; it is an amount that shows respect as a fellow chef and for interrupting their work and causing them paperwork. Maybe it's $50 or $100 if you are in deep. At times $20 will do if you live up the street, it is only a few items, and you may have the opportunity to buy them a beer sometime. Resorts are well stocked and usually take pride in their products, as it is a huge part of their guests' experience.

Another huge part of the Caribbean end game is transportation. There are a finite number of taxis available on any island. Most people are taught to follow rules in public and to wait your turn. That could get you fired if you believe our currency is time first, then money. For example, if you are on a mother ship to a sailboat in a large regatta and all the mother ships, let's say 70, hit a new island needing supplies, the island may have 15 taxis. Who gets them? You are all on a tough schedule, and even if you double up what do the rest do? Locals will usually say, "We only have so many, so good luck."

But anyone with a car can be a taxi. Talk is cheap; hold cash in the air and ask, "Who wants to be my taxi?" A fast $50 means a lot to the average islander. Booze is cheap in the Caribbean, and $50 buys a lot of drinks. $50 doesn't work? Even I will be your taxi for $100; just make sure the deal is round trip. They will probably even help you with the groceries. You may get good local inside information that a licensed taxi driver wouldn't give. Use common sense though; if you are a petite beauty, don't jump in a drug dealer's car—double up with another chef. Normally, you should beware of gypsy cabs (unlicensed civilians soliciting rides), as some are dangerous. If you ask someone to be a cab in front of your peers, there is a lot more accountability. Everyone will already be talking about how clever you are, and they will be more likely to remember the person who gave you the ride. All that being said, I always find it better to rent a car and do your own driving; there are no time constraints, and it is usually cheaper. You can also stop and do some personal errands without a meter running.

If you don't know the island and it is your first day shopping, take a cab and ask a lot of questions. I always try to scan the area so I know where to find any and everything. The chef is the team member out on the road the most, and it helps the whole team if you are a great resource. If you are in one location for a while, you may be able to

befriend some of the local fishermen. This can be a valuable relationship, but do not count on them as the vast majority of fish is shipped in.
Always remember that your peers are your best resource. Talk to them
before you go to new islands.

When you are going island hopping, remember that each island is
owned by a mother country. Each island will use one or more currencies, and some banks may not take your country's credit card. If you
work on yachts, it is not unusual to have five different currencies in
your wallet by mid-charter season.

St. Martin is an island split down the middle and shared peacefully by
two nations, Holland and France, respectively. Each side has its own
government, police, language, and currencies. They both take the Euro
and American dollar; however, the Dutch side also takes guilders, and
you can probably even pay with Eastern Caribbean dollars with the
likeness of Queen Elizabeth on the front. None of these currencies are
exactly one to one, so you will need to know the exchange rates for all.
Write them down and keep them in your wallet. Not knowing can cost
you a lot over a season, and keep in mind the exchange rate and the fee.
Most chefs are given advanced purchase allowances (APAs) in one currency. You must factor in the math on both sides when you exchange
money, as most of us will be responsible for our own exchanges. Big-
program chefs can exchange and spend hundreds of thousands of dollars in a busy season. Not doing your homework can cost the program
thousands of dollars if the chef is ignorant in this area.

Changing languages, currencies, and types of transportation, sometimes all in one day, is a hallmark of Caribbean travel. It is also part of
what makes us the James Bonds of cooking.

CHAPTER 2

The Team

So you show up on the other side of the world for your new gig. You arrive at a massive waterfront estate with a huge yacht docked out back. There is a heliport in the distance with multiple choppers and what looks like a small village at the edge of the compound. You know what your task is—you are the chef—but to whom do you report? What is your rank and place in this sea of people? Where do you get work-related monetary instruments? How about your personal bedding and toiletries? Who is your liaison to the owner? Who works for you? Who maintains your staff car, and who sets the rules for property? Will this all change when you leave on the yacht or fly off to another of the boss's properties?

Historically, the super wealthy followed the British Empire's lead and had a main core staff that traveled with them between assets: butler, valet, upper maid, chef, and sous chef. The support staff for each asset stayed in place. These would be the estate managers, the gardener and groundsmen, housemen, chambermaids, or the yacht's operating crew.

In the late 1990s, the proliferation of the Internet and cell phones coupled with advances in private aviation changed everything for private service. Lead times between the boss's initial call to action and arrival have gone from weeks or days to hours. Decisions can now be made on a whim, and the boss can wake up in Florida and decide to ski in Aspen that afternoon. Their pilot and plane will instantly jump to life and get the boss en route while the ski chalet staff light the fires and chill the champagne.

People rarely travel on their yachts when going any distance; the yacht has to be made seaworthy. Everything that is not permanently secured in place is taken down and wrapped in towels. Seating is placed backward with throw pillows wedged between them, and the dining table and chairs are shrink-wrapped in place together with a few trips around the table. The crew runs 24 hours a day delivering the vessel. Once the boat arrives in a new port, the plane is not far behind. Having a small army of servants drop everything and fly out to meet up in an exotic land on a few hours notice is truly not at all feasible. These days each asset is almost fully staffed, and only a few assistants and department heads may move around with the boss.

Yacht service requires that the chef also be a mariner and have cross duties with the operating of the vessel, not just production of food for the crew and owner. Bringing a mono-tasking, land-based cook is not often a good idea unless you have an extreme diet or the boat will not be doing much cruising.

With all this in mind, the size of yachts and estates began to grow exponentially between 1998 and 2008. Now a billionaire needs to staff several homes, boats, and aircraft. The number of jobs created was unprecedented. A super consumer can now employ between 500 and 1,000 people who have nothing to do with the boss earning a living—all of them working private maintaining or running the boss' assets or creating a comfortable lifestyle for them. If you work for a super consumer, where do you fit in this sea of people?

Although overwhelming at first, it can be broken down in pyramid form similar to an army.

Remember the number-one rule in private service? Never forget who pays the bills. That would be the *boss*. You could work for the boss for years and never properly be introduced, but your whole existence will revolve around their wishes. It is all about them, and you are being well paid for that. The boss's words directly to you trump all other instructions or orders.

Next, if you work for royalty or in a palace for a sheik or sultan, you may have what is called a *major domo*. This is a proxy for the boss at the highest level. The major domo (domestic) is the ultimate end all for the staff. Treat them like you would the boss: they are considered part of the owner's working party and may even dine with them in some circumstances. The royal or titled boss will not usually talk to staff directly about anything pertaining to work. They may have a polite conversation that they control, or if you are lucky you may receive a compliment on recent events. However, if you bring up the job at hand you will probably be looking for a new employer. If word gets out, you will probably find it hard to secure another position.

In some old-fashioned households, *the Mrs.* will take on the job of running the household and become the chief executive of the staff. This relationship puts her with one foot in both worlds. Communication usually improves, being that it is direct, but you have to have your ducks in order at all times as she will consider your department part of her turf and she will have expectations of you not only as a consumer but as part of her team. Your relationship will be a little more relaxed, and you can definitely talk about the situation at hand. Just remember, though, it's her money and her reputation on the line. Department heads report directly to her at appropriate times. This was common in America during the Gilded Age and in some old-money homes with established matriarchs who did advanced entertaining.

The *personal assistant* is someone who works with the owner everyday directly. They do not work directly on staff; however, they carry out the personal organizing of the boss's travel and social calendar and may assist in organizing events. They may be involved with ordering and sending gifts, invitations, and the boss's personal purchases. A personal assistant also has a foot on both sides of the fence. They have the boss's ear all of the time. You will probably outearn them and outrank them, but nothing is more powerful than the intimate relationship and open communication that they share with the boss every day. It is in your best interest to treat them as a member of the boss's party.

Head nurses run their show. You will typically find them on staff full-time when you have a sick or elderly boss. They report directly to the boss or the boss's physician. All other nurses and subordinate medical staff would report to the head nurse. They, too, have a foot in both worlds.

Nannies and *au pairs* take care of the raising of young children. An au pair is usually a girl fresh out of high school that trades room, board, a small stipend, and use of a car for babysitting and feeding preschool age children. Nannies are professional child raisers, and there are standards and professional societies for them. They work directly with the boss and often eat and travel with them everywhere the children may go. Nannies are best thought of as part of the boss's party.

THE DEPARTMENT HEADS

The next level of staff is in charge of specific areas of the boss's life, home, and safety. Each department head has a direct relationship with the person in charge, be it the boss, major domo, yacht captain, or in some cases chief butler. Department heads have their own staff, funding, and transportation. Basically they run their own small business with accounting and the ability to hire and fire in some cases. If you have an issue with another employee, you should speak to their department head about it. Do not pull rank; just go over their head peer to peer. Department heads can be very territorial about their staff, and each department has its own leader, staff, and politics. Depending on the consumption habits of your boss, you may or may not have staff under you. You may also have multiple responsibilities. I, hired as the chef, have also been the estate manager, chauffer, mate, steward, and purser on certain jobs. A billionaire may have 3 or 300 employees. The tasks may be spread over a few or many department to-do lists.

The *butler*, in a formal home, is in command of the staff and your liaison to the family. They can be thought of as the captain of the household. Depending on the household, this is a working/executive job, which may entail everything from hiring staff and social planning to polishing shoes and ironing. In the presence of a major domo, the butler would revert to just a department head. The underbutler, service staff, and housemen report to him. The chef and the butler work closely together, but remember that the butler has the boss's ear at all times.

The *valet* or gentleman's gentleman is the travel butler, who goes with the boss when visiting other estates and on extended stays in hotels. This is old-school British but still exists in forms worldwide. Their job is to be the liaison to the staff of an estate or hotel where the boss is staying, making sure that boss's standards and requirements are met and acting as a shield from unfamiliar staff.

The *estate manager* is in charge of everything that has to do with the physical plant and property. All maintenance workers, gardeners, and contractors fall within their domain. The estate managers may also work as pursers handling the distribution of working funds and local banking. They are usually in charge of staff and owner's vehicles and their maintenance. They handle local permits and any correspondence with the local government.

The *chef* (you) is in charge of catering. This includes planning, provisioning, and execution of meals. You are also in charge of kitchen maintenance such as the cleanliness and care of ranges, ovens, hood fans, refrigerators, hand tools, pots and pans, and walls and floors. The chef is a department head and works either directly with the boss or through a liaison such as the major domo, butler, or estate managers. Sous chefs, scullery maids, and dishwashers report to the chef, but keep in mind that the scullery maid may technically be under the supervision of the head housekeeper and be considered on loan at appropriate hours.

The *head housekeeper* is in charge of the interior of the house, including your quarters. Anything that has to do with cleaning, dusting, polishing, and laundry falls in their domain. When the maid staff has the house in order, you will have clean sheets and towels in your quarters already. Chef turnover can sometimes be only a few hours if your predecessor did not work out. If the head housekeeper is not available upon arrival, any upper maid staff should be able to assist with linens and toiletries. There are many types of maids in a large, formal household. Some deal only with the boss's quarters and are fiduciary. Some spend their days in the basement laundries, and a few lucky ones get to spend their time with us in the kitchen doing the scullery work of washing pots and pans and detailing the kitchen when we are finished creating our gustatory delights. The head housekeeper is in charge of their staff schedule, provisioning cleaning supplies, and accounting for their department.

The *chauffer* takes care of ground transportation for the boss. They are in charge of not only the driving and navigation for the boss but also the maintenance of the boss's private fleet of vehicles. They also act as the liaison with the vehicles manufacturers. These days, chauffeurs are no longer stodgy old men but a vital part of the security team. Many are personally armed and have military backgrounds. I have worked for people with personal armored cars that ranged from a stretch limousine to a stock-looking Jeep Cherokee. Driving techniques are learned through *offensive* driving classes. The directive was always, "When the shit hits the fan, remember that you are paid to drive, so do it!" In other words, distance from the situation at hand is imperative. They teach the driver how to move vehicles blocking your path using basic physics and that no driving rule applies to you until you are miles clear of the scene; police attention is your goal. The other side of being a chauffer is to make the boss comfortable. Courses are available on how to become an efficient chauffer.

The *chief* or *fleet pilot* is in control of all of the owner's private air force. All aircraft fall under their control except for spacecraft. Although several of our billionaires now own spacecraft, it is too new to see how this department will organize as they must also work with NASA. Billionaires may typically own multiple aircraft, and I have worked for two that owned their own airports just to manage their active aviation collections. The chief pilot will be involved with acquisitions and sales of aircraft, and they are also in charge of scheduling, maintenance, and fueling. All other pilots (jet, helicopter, prop plane, and hot air balloon), maintenance workers, and contractors report to the chief pilot. Pilots tend to live in hotels most of the time and are rarely seen at the estate but will come by to pick up and drop off packages and catering.

The *chief captain* is in charge of the boss's private navy. All aquatic vessels (motor yachts, sailing yachts, dinghies, day sailors, tenders, and submarines) fall under their domain. They are involved with vessel acquisitions and sales. All other captains report to the chief captain. They are the liaison to the boss, shoreside management companies, and charter brokers and are in charge of maintenance and insurance issues. We will cover the whole hierarchy of a superyacht in the chapter on yachts. Unlike on land, yacht captains are in a unique position because when the boat leaves the dock they are no longer department heads: they are gods on the water. I know this because captains tell me this—they tell everyone.

They can have the owner sequestered, put a gun to the chef's head, or knock out a deckhand with a shovel (all of which has been done) if they feel the owner or crew member is endangering the life of the guests or may cause the imminent peril of the vessel. When you are a chef on a yacht, the owner is your client and the captain is your boss.

The *head of security* runs a more independent operation. They are self-sufficient and go about the very serious business of defending the boss, the boss's family and guests, the boss's assets, and if there is time or resources also the staff. They may be the ones doing background checks on new hires and putting pressure on the boss's enemies to back off. They are around the property or the yacht but tend to keep to themselves and are usually paramilitary types who are used to being hyperfocused for long periods of time. They are the muscle and the armor. Feed them healthy and treat them with respect, but it is best not to get involved with them. Their job is to look out not for you but just the boss. If you feed them well, they might remember you when it comes time to decide who gets to live on a bad safety day.

The *sports franchise* officers hold a crossover position as they are both a legitimate business and a toy for the boss. There are two kinds of billionaires: those with sports franchises and those who don't have one yet. I don't care if it is a baseball, football, soccer, Formula One, America's Cup, or NASCAR, the team office will deal with the private staff all the time. Tickets for guests, tickets for us, dinner for team partners at the house, charity events on the yacht—be nice because the perks can be stratospheric.

YOU MUST ALWAYS OBSERVE
THE RULES OF PRIVATE SERVICE

1. Never forget who pays the bills.
2. PFA (pay attention).
3. Get it done, get it done right, get it done right now.

4. Every task in private service must have a beginning, middle, and an end. Do not leave anything unfinished.

5. There are no excuses in private service. Don't say I'm sorry. At the billionaire level, bosses and department heads tend to eat their young. Save your weakness for your personal time.

6. When arriving onsite for the first time, there is a break-in period, especially if you are high functioning. Be the FNG (new guy). If you are perceived as a threat to the status quo, the status quo will gang up on you.

7. Don't lie. Department heads and the boss are rarely stupid.

8. If you do something really bad, and this will happen at some point, rectify it. Do not get caught, and never tell another living soul—ever. This is not a good field for vulnerable or emotional do-gooders unless you are the nurse.

9. In volatile times, stay under the radar. Stay out of staff politics.

10. Do not be a snitch. Do this for you, not them: no one likes a snitch.

11. Do not be flashy. We are all paid vastly different amounts of money based on how much responsibility we have. You will work with some staff who are millionaires and some who are beggars. The more you show off, the more generous you will have to be when you socialize with them.

12. It is nice to be liked, but it is more important to be respected.

13. Avoid drama.

14. Keep your private life private. Leave your work at work.

15. When another team member is down or too much is put on their plate, help. What goes around comes around.

16. You are a highly paid artist with a benefactor. This is high on Maslow's list. Try not to make others feel small.

17. A high tide raises all the boats in the harbor. Nothing affects team output as much as the hiring process. Don't skimp on staff. If you can taste the savings, you have failed. Take hiring staff seriously. Every team member suffers from a bad hire.

When you first arrive to the yacht or estate, I recommend taking 20 minutes to immediately get your clothes situated and your bed and towels in order. If you are hitting the ground running, you may find yourself wanting to go to bed and find that the other staff have left or the towels are in the boss's area and end up sleeping in your clothes. Trying to dry off without a towel after your morning shower is not the way to begin your second day.

Remember that you are not on staff to make friends and that partying with most staff is a huge liability. Anything you say or do is likely not going to be left at the bar or party. I cannot tell you how shocking it is when the yacht captain, who stayed in, tells you verbatim what you said and did the night before, even though you were buying the drinks for the crew. Just because you work together does not always mean that that your fellow staff have your back.

It is easy to covet what you see every day, and this type of work can be exciting and romantic. Oftentimes, private staff are hired partially based on their appearance. If you value long-term self-preservation, please remember that you have other options in the dating pool. Do not date the staff unless it is absolutely unavoidable. I would ask myself: is this relationship worth more than this job or even my career? Sometimes it is. Some people find a mate for life on the job, one with whom they can work as a team from there on out. True love trumps all, but make sure it is not just a fun romance because when it ends it can put a huge burden on the staff. If the staff feel it, that vibe is likely to bleed over into the boss's area. The boss may not know what the problem is—just that there is one.

CHAPTER 3

The Seasons

One of the great things about being the chef to a billionaire is that you do not sit in one kitchen all year long staring at stainless steel. Every few months you usually get a change of scenery, meet new people, face a new set of challenges, and get to live and operate in some pretty exotic locations. Although working for the moneyed elite is no guarantee of this, the vast odds are in your favor.

While there are definitely four seasons in the private chef world, they do not necessarily coincide with the Christian calendar. I like to think that the year starts fresh for many private chefs at the end of October. This is due to the Ft. Lauderdale Boat Show coupled with the fact that it is toward the end of the time-off or vacation season for most private chefs. I will focus on the yacht chefs for a moment as the estate chefs move around less and at a slower pace. This is the time of the year when yacht chefs take their vacations.

For a yacht chef, most of us have finished up the summer season; the boat has gone to the shipyard for any upgrades, repairs, routine maintenance, or interior refits. During this time, many of the crew members have left due to attrition. Some were freelance or on a seasonal agreement, and some were just not a good fit. Captains do not like to pay crew salaries when they are not needed for long periods. There are day workers, who are available for hire on a short-term basis to do the dirty grunt work in the shipyard. Many of these are foreigners who are making a full-time living at this or are yacht crew between jobs. They are convenient because they are paid as needed, do not need to be housed, and are offered only a basic lunch that they eat on the dock.

Most yacht chefs earn a minimum of one month of vacation per year with pay and a round-trip plane ticket to the destination of their choice. If not employed full-time, many will save accrued vacation pay and time to be used at the end of a season and coast while looking for their next challenge. At the end of their time off, most yacht chefs will get the travel bug and migrate to Fort Lauderdale, Florida: the super yacht capital of the world.

Each year at the end of October, the largest and most vulgar display of wealth on our planet occurs along Fort Lauderdale Beach. If you have not seen it, you do not have enough letters in your alphabet to describe it. "Boat show" is such an understatement. Megayachts, superyachts, and gigayachts worth billions of dollars line the permanent and temporary docks along a stretch of over a mile of the Intracoastal Waterway. The aft ends of 200-foot long yachts are squished in together like townhouses with really good porches. Gorgeous models from all over the world line the tented pavilions and give tours of impossibly expensive

FIGURE 3.1   A MANSION IN FORT LAUDERDALE, WHERE THE YACHT IN BACK
MAY BE WORTH MORE THAN THE HOUSE.

FIGURE 3.2   SUPERYACHTS IN FORT LAUDERDALE.

FIGURE 3.3   LARRY ELLISON'S GIGAYACHT HORIZON OVERSHADOWING
LESSER BILLIONAIRES' YACHTS.

objects. Free food and drink flow everywhere. The latest versions of the most expensive, handmade supercars perch on floating docks and perfectly manicured patches of lawn next to the VIP pavilion. Billionaires roam in tight packs, surrounded by their handlers, yacht brokers, captains, and mysterious trophy beauties. To call it quite a spectacle is a total understatement.

Yacht crew and industry professionals fly in from all over the globe to see the latest and greatest floating palaces and to connect with peers. It is the only time of the year where everyone who is anyone in the yachting business can be all in one place. The show is also the time when purveyors and provisioners have the department heads as a collective to try to win our favor. The parties are epic. Have you ever been to a gathering of over 5,000 of your peers with all the free food, drink, and entertainment you can handle? How about two nights in a row? How about for a whole week where you may attend close to 20 parties? Ever been to a $500,000 white party (everyone wears only white) on a ship with bars made of ice that are 60 feet long and where brand-named Swiss watches are raffled off as door prizes? All while Russian models spin from curtains from the ceiling 30 feet above and others pour shots through an ice sculpture luge for you? Welcome to the rock star world of yachting.

But once the show is over, the parties end, and everyone's batteries are recharged, we enter into the hiring season for most of the crew. Chefs are expensive so they are usually hired last. Permanent chefs are usually hired first, followed by seasonal chefs. Freelance chefs are always hired at the last second because they are paid the most and charter clients want to wait until the last minute to hand over the advanced purchase allowance (APA) as it is usually nonrefundable. Bad weather, a global catastrophe, or a big enough dip in the stock market can cancel many trips, and the APA can be as much as six figures or more; therefore, free- lance chefs are on call to leave at a moment's notice toward the end of the hiring season, which runs from early November to mid-December.

The first of the two work seasons begins. The winter Caribbean/Bahamas season can be a crapshoot for charter yacht chefs. Many seasons have been spoiled with big winds and waves, where it was not suitable to vacation on the water. While rich people tend to vacation all over the world, billionaires tend to run in packs. The vast majority will focus most winter getaways around the tiny paradise of St. Barthélemy (commonly known as St. Barths) and Anguilla. St. Martin is the work hub for all yacht and villa provisioning on any large scale. Jumbo jets arrive all day from France, Holland, and America, so you can get almost anything overnight if you are willing to pay for it.

Roman Abramowitz and other super consumers can park their private 767 jets and enjoy a 10-minute limo ride to their gigayachts, some of which are color coordinated to their planes. St. Martin is where yacht crews live between their jobs. The island features the best nightlife in

FIGURE 3.4 YACHTS ON NANTUCKET ISLAND IN MASSACHUSETTS.

FIGURE 3.5 SIMPSON BAY IN ST. MARTIN: THE CARIBBEAN CAPITAL OF YACHTING.

the Caribbean, and crew live on a grand scale there. I have watched crew fall in love and some die there, but it is never boring.

If your boss is renting a villa in the Caribbean, chances are your foods will be provisioned through St. Martin or St. Thomas. If you are Bahamas bound, your foodstuffs will most likely come from Nassau. The Bahamas has come a long way in the last 10 years as far as provisioning goes. It used to be tough to get what you want there, but now it's set up for business. Small planes or mail boats can send out goods to the smaller islands from Nassau. Items coming in from the States can be opened at customs and left on the docks to spoil, so it is better to pay the price for professional provisioners than lose the food on your way to an underdeveloped region.

South Florida is also a seasonal hotspot. Palm Beach may have the largest collection of millionaires. Despite its season being only 10 weeks, it is littered with major events, and chefs compete to make their boss look good. Although not everyone can have the biggest mansion, anyone with a big enough seasonal budget can have the best food and service.

When you combine a competitive and social boss with enough resources, you really can make some magic happen, and it feels good to run on all cylinders for a while. Palm Beach is a great place to work. Everyone is there to have a nice time, and even the grocery store has valet parking. The town is not necessarily famous for its restaurants because all the best chefs are working privately and the big players mostly dine at each other's homes. Nightspot staffs tend to remember your name and who you are working for, and they generally treat you with respect. It is a happy town with a lot of fashionable people. The homes are unforgettable for their design and scale. Miami is also a hip place for villa rentals and short-term yacht charters. Both cities still provision in Fort Lauderdale as it is better set up for this business and nicely placed between the two locations.

February and March are the strong charter months in the Caribbean for both yachts and villas. The reason is that many schools have vacation periods during this time and many families have tired of skiing and are looking for fun in the sun. Once April 1 comes around, there is a mass exodus out of the Caribbean. The cost of departing plane tickets skyrockets. A few major events close out the season and create massive demand.

There are also nice events in Fort Lauderdale for returning yacht crew, and when we are back the boats go into the shipyards and marinas to go through postseason maintenance. During the exciting and exotic Caribbean season many crew have fallen in love and now want to work on a new boat as a couple or as a "team" as we call them. This is a better idea than having your new lover join your boat. The politics and crew dynamic are different if you are hired as a team. I usually don't acknowledge a couple unless there is a ring on the finger and it is serious. Just because two 25-year-olds want to sleep together and travel the world together does not mean that they should get their own room and special privileges. There are exceptions, but if the captain hires them as a team you have to respect the captain's decision. Others now have another season under their belt, and it's time to move up in the size of program or to go to another program to take on more responsibility. Sometimes it's just that you have gotten all you can out of that experience. Yachts turn over crew for many reasons, but they do it a lot more than any job on land. This is likely because we don't go home at night, we live in tight quarters, and it is a 24/7 lifestyle job.

Springtime is wonderful in South Florida; there is lots going on and great weather. At this time, however, the satellite offices start to come to life in Europe. Antibes, France and Palma, Mallorca are the yachting capitals of the Mediterranean. Sure, there are other areas like Monaco and Viareggi, but if you are crew and have to spend the last of your savings to go after a yacht job, you are going to fly to Antibes or Palma. Both are fantastic locales.

In the summer, most of your billionaire bosses are going to use their assets in the Mediterranean or the northeastern United States. Newport, Rhode Island, is the summer yachting capital of the United States, and the Hamptons are ground zero for estates. As amazing as the Mediterranean is, it is hard to argue with the seasonal quality of life on Nantucket Island or the action in a celebrity-filled Hamptons nightclub. Both the northeastern United States and the Mediterranean are stunning and ready for action. You should consider yourself lucky to do both during your career. Having a fantastic meal and a brilliant scotch, seeing a famous band in an old barn, then walking back to the fogged-in docks down cobblestone streets is the kind of heaven Nantucket is famous for. Hopping countries and cultures while changing languages and cuisines and never having to unpack is beyond what you can imagine and what makes the Mediterranean so special. I probably should mention that southern California has the most private estate chefs. They tend to work year round and are not seasonal.

After Labor Day the yachts and villas run down their stores, and chefs migrate back to their home ports and cities. Now chefs enjoy their well-earned time off. It is only a matter of time before you get the bug again and start thinking about the Fort Lauderdale Boat Show or your friends back in Palm Beach. I realize that during the working seasons for your bosses you may live in Montecito, Manhattan, or Dubai, but the fun seasons for your bosses will be the work seasons for you.

CHAPTER 4

Starting a New Gig

OK, so you have scored that new dream job. You fly to a destination somewhere in paradise, well dressed and ready to make a good first impression. Upon getting out of the cab, you are greeted and—now what? There are a lot of instructions, but as mentioned earlier I highly recommend getting yourself personally situated first. Check your accommodations and turn in any travel receipts right off. The sooner you turn them in, the sooner you get reimbursed. The general rule is that you get yourself to the airport on your home side and the boss picks up travel expenses from there on out. The boss's party should arrange for airfare. Not only is it their responsibility, but it also begins to form a paper trail between you and them and a record of the starting date in case of any confusion down the road. The vast amount of private chefs jobs are begun on an oral agreement or handshake. Contracts are usually signed only after a trial period.

Remember, be polite and embrace the FNG status for as long as necessary.

Most jobs will start with a tour of the property or vessel. Don't sweat the details; often a billionaire's assets are dazzling and complex. You will have time to explore on your own time. The most important characteristic of a successful private chef is being a good detective. I casually ask questions of everyone along the way. Sometimes people are helpful and they tell you what the last chef (who may or may not have been successful) did without realizing that the owners did not enjoy the methods or choices and just were not vocal about it. If you keep hearing the same information again and again, it may be good information. Usually there will be notes made on the boss's preferences. Keep in mind that oftentimes the information on them may be out of date as diets and trends tend to change or evolve.

Frequently, you will be brought to a property or vessel well in advance of the boss's arrival. This is a luxury. You will have time to clean, organize, and do an inventory at a more reasonable pace. A good starting point is a thorough cleaning. I take everything out of every drawer and cupboard. You will need to do an inventory of the equipment, even if just mentally, and make sure that you are not missing anything crucial and that everything is in good repair. Even if your workspace already appears detailed, it gives you an intimate view of your assets, you will be confident that everything is sanitary, and it shows the other staff members that you are serious, detailed, and not lazy. This simple act sends a huge message that usually makes it all the way to the top of the organization.

Staffs love to talk and gossip, and anything you say can or will be used against you at any time. Smile a lot, but keep the humor under control as it may not be appreciated throughout all the varied cultures and ranks. Bonding with a new team is a real art. Most private staffs are a melting pot of cultures, genders, and sexual orientations, so it is best to try to fit in and not try to get the team to have to adjust to you. Be most careful when initially socializing with your new team. Think of them as respected family members, and dating any of them should be considered taboo.

It is customary to surrender your passport to the captain or the boss as symbolic collateral for use of the boss's monetary assets and vehicles. They will feel more comfortable that you are not going to steal the Range Rover, drain the AmEx Black, and run away with the $10,000 cash they gave you to provision with. I have never heard of anyone not getting their passport back, and if you need to do a personal trip you will have to turn in those assets before you leave anyway.

At times a new asset is the reason a new chef is hired. This can be fun, as you will now have a blank canvas to work with. Outfitting a new kitchen/galley is both an awesome and daunting task, but what a feeling to start with everything new. Choosing knives, pots, pans, cutting boards, appliances: it's like when a record company gives a rock star an advance for new guitars to tour with. It is very important to get this right. If the kitchen/galley is out of sight and more of a staff workspace, then you may not want to buy everything that is over-the-top, show quality. If the space is part of the main living area and decorators have spent hundreds of thousands of dollars on cabinets and tens of thousands on the ranges and refrigerators, then you may not want industrial equipment lying around draining the sex appeal of a fine designer kitchen/galley. You must take into account the space, the standards of the boss, and any budgetary requirements. It is crucial not to let your ego dictate that this is yours and you are going to be this boss's chef forever. You may love copper cookware. Copper cookware is the most expensive, but it is a lifetime investment (so technically it is the best investment)—but only if you own it forever. It also not only requires cleaning but also a chemical polish, two distinct processes. If you do not have a scullery maid, it is double the work. Copper cookware is not the best choice for amateurs. It is like a Lamborghini: it is a car and is high performing, but not everyone is going to know how to drive or maintain it. Consider anyone else who ever has to cook there when you are not or who will be replacing you down the line. If you are a classically trained chef hired at top dollar to do Michelin star food, maybe copper is the answer. If not, it's best to wait six months to put in the big requests. The moral of the story is to match the equipment to the space to the way the boss likes to eat. At the end of the day, you are not there to teach the owner new things (unless requested), and you are not there to prove to the boss how awesome you are; it has to be about the boss, not you. What do they like? What standards do they have? How much money do they want to invest in their kitchen/galley? Keep in mind that you do not have to get every little gadget right off. It is best to avoid sticker shock at all costs. Even if you are going to use everything eventually and it is easier to pick it all up at once, big spending can spook the most extravagant billionaire, estate manager, or yacht captain. Perception is reality, and that goes a long way in earning their trust.

Many times you will arrive to a new job either midstream replacing another chef or fly in hours before the owner does. You will literally hit the ground running. Your adrenalin will pump; decisions are made on the fly and by instinct. Every second will count, and there won't be time for tours or meet-and-greets. Get your stuff in your room, and stay focused on that first meal. Once the meats are thawing, dessert is under way, and breads are rising, you can take a quick inventory and see what you need to get through the next 24–48 hours. If you can make it out to pick up what is on that list, do so posthaste. If there is a team member who can shop for you, stay and work on the task at hand; however, this will rarely be the case. After the dinner is complete you can take an inventory and make plans to shop after the breakfast the following morning or have things delivered. If you are not familiar with the area, do your homework online before you get on the plane. Also realize that your peers are your best assets. Seek advice from other chefs who have been in the area recently, as businesses in resort and luxury areas do not always stay in business indefinitely. Having many of your peers on Facebook is helpful because you can put out a bat signal for information. Most chefs like to share their knowledge and experience. Often they can give you the business owners' names and contacts in advance. Once again, being a good detective at all levels is the hallmark of a great private chef.

Eventually you are going to meet the boss. Billionaires like to meet their own private artists, and the better the communication, the more success you will have. Once again, dress to impress and be polite. Wait for the boss to offer their hand; however, some do not like to shake. A simple how do you do and a smile will go a long way. People tend to mirror to you the way you come across, so be positive, be brief in your replies, and know that when you are in the boss's section of the estate or yacht it is not the time to have deep and meaningful conversations. If the boss comes to your work area, it usually means they want to talk to you.

The best way to make a good first impression on your first day is to be polite, be competent, stay focused, know your place, and remember why you are there. Private chef jobs are lifestyle jobs. At no one time are you doing too much heavy lifting: the workload is spread out over many hours, and it is more of an endurance race. You tend to switch

FIGURE 4.1   CHEF NEAL IN GALLEY OF ONE OF THE MOST FAMOUS YACHTS
OF THE MILLENNIUM.

between work, play, and home life throughout the day, and most chef positions at the billionaire level are live-in. You may genuinely enjoy and socialize with your fellow staff and crew mates, but remember that you are hired to do a challenging job as a highly paid professional, as are many of the staff members surrounding you. Being a good staff and crew mate is just as important as being good at your individual job. If you are the best chef but a pain in the ass to live with or socialize with, you will not last, and if somehow you do the people/victims around you won't. Turnover is inefficient to your boss. Causing unnecessary turnover should be considered stealing from your boss.

Being a private chef at this level is truly an amazing experience, and there is a real James Bond–like feeling when you are on the plane ready to take off to some exotic location to join an elite team of professionals to do something amazing for one of the world's most demanding clients. When you arrive at the guard gates or the marina, everyone knows who you are and is expecting you. People carry your bags and look at you like you are someone special. The stimulation of going somewhere exclusive and having the privilege of working in or on a masterpiece of engineering or design as a trusted, valued professional is not something everyone gets to experience.

It feels good to join a new program filled with an interesting cast of characters working seamlessly together like a well-oiled machine. At the opposite end of the spectrum, you may climb aboard a literally sinking ship. Every yacht that sank had someone cooking onboard, and I have known a few. Whatever the circumstances, the adventure begins.

CHAPTER 5

The Food and Techniques

As we begin to head into the actual food and how-to areas of this book, professional chefs may find these subjects and offerings familiar territory. I believe it is important for everyone to be on the same page with the terms and techniques. There are many styles of cooking and specialized techniques to make food fun and interesting, but when your ingredients are the stars, restraint and rock-solid rudimentary skills are required. Because private chefs do not sell food, there is no reason to add too many distractions to a supreme piece of meat or fish. We rarely do any wet marinating, and most of us do not make a common practice of serving "dishes". No chicken tetrazzini or angels on horseback. These are best saved for staff meals where we hide discount proteins behind cheaper ingredients.

Fine gourmet or Michelin star quality cooking focuses on creating balanced plates of a protein—usually of no more than four ounces or the size of a deck of cards. The protein should not have any gristle, bones (barring a proper chop), or anything that would stick in the teeth or come back on the plate after the meal is finished. The protein is minimally seasoned with kosher salt and pepper and then perfectly cooked and paired with a great sauce on the side. The plate is finished with an

FIGURE 5.1 BRUNOISE OF VEGETABLES.

appropriate starch and flanked by two vegetables: one green and one of a colored variety. Seldom do we do pan sauces unless it is a pasta.

The theory is that if nothing in nature tastes better than naturally concentrated beef or chicken flavor, why water that down? I am not talking about the commercially raised beef and chicken that you are used to from your local grocery store. I am talking about naturally raised, prized heirloom animals, bred for their marbling and flavor.

Grocery store meat looks plump and wet, and it looks appealing to the untrained eye. But most animals from which this meat comes are fed diets that make them retain water. Food is sold by the pound, and if the meat is wet it is heavier and cheaper to produce; it's simple economics. That type of food is chewy and can also plump you up.

Top-quality chicken is not wet and slimy. It is not bright and shiny and does not leave liquid behind in the package. It is concentrated in flavor, tastes absolutely delicious, and is also expensive. I once had 12 poulet du Bresse flown into St. Martin overnight from Paris, and it was US $1,100. The chickens still had their heads, feet, and innards. Only the feathers were removed, and there were pebbles in their digestive tracts. They were tough to butcher and harder to cook, but the concentrated, almost sweet flavor made it worth the while. The flesh is tight and really sticks to the bone when raw, and the meat is sweet

FIGURE 5.2   NEAL HOLDING A POULET DE BRESSE IN ST. MARTIN.

and tender when cooked. The fat is a rich, deep yellow that is perfect for rendering.

At some point, most classically trained chefs stop learning the new trends and tricks and become custodians of the enduring quality products and methods of working with them. It is not so important to know about all foods, only the best foods. It is important for us not to lose the demand and supply for these rare ingredients and the disciplines and techniques on how to cook them. This discipline is not for everyone outside of the private chef world as it is hard to make a profit off such costly products. Not only the protein itself but also this quality of

FIGURE 5.3   GORGEOUS SIDE OF PATAGONIAN TOOTHFISH.

ingredient demands that all accouterments be up to snuff. French butters, Italian white truffles, and $250/gallon Greek olive oils are appropriate here, and anything less would be like putting snow tires on a Ferrari. A chicken dish selling for $100 per plate in a restaurant is not likely, but this can certainly happen in the private chef world.

If you are going to be the chef to a billionaire, you are going to need to be knowledgeable about ingredients that the vast majority of chefs will never actually see in person during their lifetime. This is not to say that all billionaires eat this way, but at some point most will want to impress or show respect to someone—like someone who may know the difference between a good fur coat and a bad fur coat.

Let's take into consideration prime, dry-aged beef. When I tell the average person how much we pay for some of the meats we use, I am often met with guffaws and exclamations that I am being ripped off. They often brag of the quality of the steak that they buy at their local chain grocery. It is hard to win an argument with someone who has never even seen this level of meat. It would be like trying to explain an F-18 fighter jet to the Wright brothers.

Let's start at the start. Like an Olympic athlete, only about 2% of the population has the natural conformation and genes to be considered for entering a special diet and training program to work toward becoming an Olympic athlete. Similarly, only 2% of cattle have what it takes to be considered to enter the diet and training program to become prime cattle. Prime is the top grade of beef in the voluntary USDA grading system. It is judged by the marbling within the meat itself, not on the outside. To achieve prime grade is no accident; however, an average cattle cannot achieve it regardless of the diet. Heirloom breeding, not unlike in horse racing, determines the potential. Just because you are a cow with great parents does not mean you automatically make the cut. In fact, if you are lucky enough to be a perfect cow, you don't make it either: sorry, but it's a boys' club. Only perfect steers are used, and there really is a big difference.

OK, so you're a perfect steer; now what? Next, you will be sent off to special camp and fed an expensive diet. Beer and massages are rumored for the Japanese steer. I have never been to Japan, and there are strict export laws for true Kobe Wagyu (Kobe is the region, and Wagyu means cow), so I, like most chefs, have never seen real Kobe. I have only seen Kobe-style, raised at Snake River Farm in the United States, and a German-owned, Polish-raised version. There is also a farm in Australia that claims to be related. All are excellent but not the genuine article.

The steers that enter the prime program are not only more expensive to buy and house (they are not jammed into pens) but are also kept alive an average of six months longer than the average steer, all while on that expensive diet. When it becomes time to harvest the animal, humane slaughtering is incredibly important. Oftentimes the steers are exposed only to a small group of friendly ranchers for the last 30 days to build a familiar trust, so as they guide them to the slaughter the steers do not experience stress. If the animal is not relaxed at the time of slaughter or if the slaughter is not done cleanly and on the first pass, the steer will panic and pump adrenalin through the musculature and taint the meat. In addition, the carcass will not drain properly and will retain blood, which later turns to a dark red, jelly in the meat. In 10 seconds, you can turn a $40,000 steer into a $4,000 steer. Losses are factored into the end price for the successful slaughters.

OK, so you have successfully chosen a good candidate, raised him like a son, and killed him with compassion, and he has been broken down to marketing forms. Is there more? Yes.

Now the magic happens. You see, up until now everything you have done has been to create a spider web of soft fat evenly marbled through the meat. But we are far from done. Prime meat by nature is drier and smaller and carries a larger fat cap, so the yield is already less to start. Now comes the aging, and aging is done with all beef. "Fresh" beef is a myth. It takes time for the meat to break down, tenderize, and mellow in flavor. There are two types of aging: wet and dry. Wet-aged beef is cryovaced and aged in its own juices. Remember, all food is sold by the pound, so wet is heavier and makes for better commerce. Also, if the leaner, more watery beef dries out, it shrivels and turns to sawdust when cooked. Dry aging is done when you have raised a beautiful animal that has a marbled healthy musculature. It is put totally exposed into special rooms with filtered air above the safe temperature zone at roughly 40–42°F for 20–42 days. During this time the outside basically turns to beef jerky and will have to be cut away later (smaller yield again). The inside will have lost up to 10% of its water weight by evaporation on top of all other lost yield. The fibers will have decomposed just enough to weaken them and take on a nutty flavor. Once cut into its final portion you can look at the cross section and see the gorgeous bright white spider web marbling on a deep red canvas. Once dry aged, we wrap the meat only in paper. It has to be able to breathe, and it is dry enough not to leak.

All prime, dry-aged meat is gorgeous. But just like there is a difference between a good fur coat and a bad fur coat, there is a difference between a good prime, dry-aged steak and a bad prime, dry-aged steak. Abundance of marbling is a guarantee. It is not judged and marked prime without it, but if the marbling is too thin there won't be enough fat to lubricate the meat. If the marbling is too thick the fat won't completely melt and will be awkward to chew. The coloring can also give away whether the meat was aged for 20 or 40 days. Time, rarity, feed, labor, loss, equipment, and transport add up to one expensive piece of meat. I have seen certain cuts run between $20 and $70 per pound plus shipping. It's like a doctor or a lawyer: they are all in an elite club, but they are not all good or the best.

Your job representing a billionaire's standards is to know the difference and not overpay for it. That steak dinner on your boss's $80 million yacht should be better than what they can get in port.

FIGURE 5.4   LOBEL'S PRIME DRY-AGED SHELL STEAK AND PRIME TENDERLOIN;
CRYOVACING ON THE SHELL STEAK DONE ONLY AT AND FOR THE SHIPPING PROCESS.

FIGURE 5.5   LOBEL'S PRIME DRY-AGED SHELL STEAK, ARGUABLY THE FINEST STEAK IN
THE WORLD. THE CRYOVACING PROCESS LEGALLY HAS TO BE DONE DURING SHIPPING.
WHEN I PICK THEM UP IN PERSON, THEY ARE WRAPPED IN WHITE BUTCHER PAPER ONLY.
NOTICE THE GORGEOUS MARBLING.

The best steaks consistently come from Lobel's, a tiny butcher shop in Manhattan. The Lobel family, who originated in Austria, has been in the cattle business since the 1840s and for five generations have been purveyors of the finest meats at their Upper East Side location. The shop deals almost exclusively with the ultra-wealthy. I have had billionaires say that they cannot afford them—meaning that they won't appreciate the difference per dollar between them and other prime butchers. If they know you—and more importantly if they know your boss—you may get some of the private stock hidden in the back that would blow both your mind and your wallet. I do not vouch for shipping or billing here, but if you have royals or brand-name captains of industry at the table, I would not count on anyone else for this level of quality.

You never feel more alive than when you have $2,000 worth of shell steak on the grill that will be perfect in minutes and ruined shortly after. Once cooled and rested, it—like great coffee, scones, or fresh orange juice—is really only good for about 30 minutes. It is not the difficulty of the task; it is the focus, timing, and discipline it takes to nail it within the window of perfection.

FIGURE 5.6   MEAT CASE AT LOBEL'S.

FIGURE 5.7   Lobel's shell steak resting before it is sliced for plating.

I have focused on the beef here to show you how much you need to understand about one protein that may only be one of seven on a buffet with dozens of sides and sauces. It is enough to scare an amateur out of even trying the experience. You must take the time to learn about the qualities of superfine ingredients. It does not have to happen all at once. I recommend shopping when you are not in need of anything. Get to know the purveyors and ask, Google, and research them when you are not under pressure. It is one of the best parts of the job. There are many sources out there for learning to cook and the theories behind the techniques. I will try to simplify the techniques to what is going through my head while properly cooking.

There are only a few pure methods of cooking, and there are many subgroups and combinations. Equipment, styles, and ingredients vary greatly, however; whether it is a copper sauté pan, a cast iron skillet, or a 40,000 BTU wok, frying is frying, sautéing is sautéing, and pan searing is pan searing.

Let's start with the pretreatment of proteins. When I am talking about proteins here, I am talking about meats, fish, and poultry. There is a lot of misleading information out there. It is true that if proteins are exposed to temperatures over 40°F, bacteria start to grow and double at intervals. The fact is, to achieve truly succulent meats that are pink from edge to edge or fish that will swell and be teeming with that burst of FOA (fresh ocean awesomeness), you need to start with your proteins at room temperature.

Heat makes things expand, and cold makes things contract and condense. If you start with a cold steak, you can sear the outside fine. However, it will take longer, and you will begin to overcook the layer just under the crust. By the time you have cooked the center to medium rare, it will be well-done two-thirds of the way through. To achieve room temperature without risking food-borne illness, you need to take into consideration the quality of ingredient, the storage, and the environment in which it is being prepared.

If you are talking about the finest ingredients stored in your subzero fridge and taken out to temper just before a meal in your custom, granite kitchen in your fine, climate-controlled home, you are probably going to be okay. We bring the proteins up to temp quickly or in an airtight packaging, such as a zip-top bag with the air squeezed out or on a platter stretched and sealed in plastic wrap. If frozen and cryovaced, you can bring them to temp quickly in warm (yes, warm) water because without air bacteria grow very slowly. Once the protein is up to temperature, let it breathe for a little while. Just prior to cooking, it is time to season. Kosher salt (a mild absorbent form of sea salt) or straight sea salt is the first line of defense in a war on bland food. The salt leaches out the top layer of juices quickly, which when exposed to high heat caramelizes quickly and creates an intense, superconcentrated crust that works as a moisture barrier keeping the rest of the juices trapped inside. The heat that passes through the food heats the juices and internal fats and makes them expand and swell. This puts stress on the fibers and weakens them, making them tenderer. It also allows the juices to be pushed outward and to flow around the protein keeping the inside consistent in texture and equally moist side to side.

When seasoning a protein, you must first season it with the salt and a pepper of choice to straight cook succulent, well-balanced proteins. After that, additional seasoning and methods are up to the chef's discretion. Most classically trained chefs who do any kind of volume keep on hand a mix of 80% kosher salt and 20% butcher's block black pepper. Many will say, "Why not use fresh cracked pepper"? The answer is that when doing any moderate volume of cooking, with a pepper mill you will probably end up with carpal tunnel syndrome, it will slow down the cooking process, and it will add all kinds of bacteria to the pepper mill when cooking in a hot environment on the fly. The truth is that salt is a preservative and the salt dust will surround the precracked pepper and hold in the freshness for quite a while with minimal loss. If you taste the kosher salt or pepper on their own, they are too strong to the palate, but at 80/20 they are a pleasing balanced blend. You will also be able to season more evenly around the protein and your process will be faster. Everybody wins!

Do not use table salt in cooking as it is a rock salt and is highly concentrated and you can't use enough to do the leaching. It also doesn't break down easily, and you can taste the salt more than the product. Use table salt at the table when you need a drastic adjustment to an underseasoned protein. Do not use kosher salt at the table, as it is weak and too textured.

Once you have seasoned the room temperature protein, it is time to decide if it needs an external alternative fat source to its own fat cap or internal marbling. Shell (New York strip) steaks, rib eyes, and veal chops do not need oil or butter on them before grilling, but they do need it in pan searing. Tenderloins always need an external fat source, such as olive oil brushed over it on the way to the grill or in a pan on the flat top. Okay: the protein is ready, and it's go time!

Here is what goes on in my head when cooking in the proper master methods.

Roasting is the highest form of cooking, the way kings and queens ate for thousands of years until modern cooking. Roasting is great for buffets and celebrations, but great roasters are born, not made. An innate, sympathetic understanding of what is going on inside of the joint (larger muscle), bird, or fish while cooking is needed. When roasting, preheat the oven or spit, bring the protein to room temperature (within reason if the joint is very large such as in a steamship round), season, and decide if the protein needs an alternative fat source. Then it is time to sear.

Searing can be done via the flat top and the oven methods. Smaller cuts of meats will be seared on the stovetop in a pan with a good amount of oil until the whole surface has a good crust. You can roll the meat around on all sides and then place it on a rack, or you can put aromatic vegetables on the bottom of the pan to soak up juices and be used in gravy later if desired. For larger joints such as a prime rib, we start the oven at 500°F and place the joint on a rack and straight into the hot oven. After 10 to 20 minutes, depending on the size of the joint, we then turn the oven down to the stable roasting temperature, maybe 350°F. Next, there are two types of methods to keeping the top from drying out. The first is to pour juices or clarified butter over the top periodically during cooking. The bulk of the juices evaporate at 212°F, so if you baste with juices you are cooling the top. If you baste with butter, which is a fat that traps heat like a bank and can reach much higher temperatures, you are using an accelerant to crisp the top. The other method is gravity basting, in which heat rises and your 350°F oven can actually be many temperatures in many places. A fan or convection process can help even this out, but the bottom of your protein is still going to cook slower than the top. To get that perfect roasted chicken or turkey requires you to turn the bird or roast periodically. Take that perfectly crispy top and put it face down in the steam coming off those beautiful aromatic vegetables swimming in the juices below. Let that underdone bottom see some glory facing up for a while. Then turn it again and again, with the show side having the last upright session.

Once the meat is almost done, pull it and let it rest on top of the counter on a rack and platter until it's time to cut. Resting is so important because you need to let the energy in the protein finish the cooking from the inside with no extra external energy put into it. I do not cover with foil as it promotes moisture at the top, which you have worked hard to make crispy. When it is rested and time to cut, you can always flash the roast for a couple of minutes in the oven just prior to slicing. This is okay only once it is fully rested. It is then sliced diagonally against the grain to weaken it and make it the easiest to chew and digest. I never use a thermometer. A born roaster learns to use one like training wheels, and then once they have the hang of it the thermometer goes by the wayside. At one end of the spectrum, the higher the temperature, the more you have to watch it and the less room for error. However, it does not take two days in a slow oven to make a tender turkey. That's for scared amateurs, and you would ruin a world-class bird that way.

You see, once you choose the method of roasting, the protein is read and reacted to. Decisions are made intelligently with roasts; it is not one size fits all, and following a recipe does not work as ovens vary so much as well as the temperature of the meat and its specific internal qualities. I suggest reading all you can on roasting, practicing, and then throwing the books away and reading the protein itself.

Pan searing is a subgroup of roasting and can be thought of as a single-serving roast. It is the cupcake of the protein world. In modern times, pan searing has earned its own place in the master cooking method hierarchy. On yachts and estates, you will do more pan searing than any other method of cooking. You will always start with your room temperature, seasoned protein, then add your liquid fat to a hot pan, and then place the show side of the protein down. If it were a rack of lamb or breast of duck, you would put it fat-side down. As soon as the protein is doré (golden brown), you turn the protein and place the whole pan into directly into a 400°F oven. Once again, you read the protein to determine doneness. Pan searing is a beautiful method of cooking as it takes advantage of both efficient and soft heat. The fat in the hot pan is very efficient; the heat of the oven is soft and gentle. Stick your hand in a 400°F oven, and you will say, "Oh, that's hot; better not stick around there for long." This is soft heat. Stick your hand in a 400°F Fryolator, and you're going to have a bad safety day. This is efficient heat. The pan will slowly sear the plate side of the protein in the oven while the soft oven heat will push the juices to the center from all sides where they will expand and repopulate the protein. When just before done and the protein is colorful and swollen, remove from the oven and pan and let rest on your cutting board if it is to be sliced. If not, it can go right onto a hot plate.

Sauté means "to jump" in French. It basically means to cook quickly in a small amount of fat in a pan on top of the stove. Medallions of veal and chicken cutlets are done this way. Imagine a fine escalope of veal with a pinch of kosher salt and a pinch of white pepper dredged in flour, with the excess spanked off and sautéed in whole, room-temperature French butter, a minute on each side, then off to a hot plate with a morel demi-glace dancing with the warm ghost of a flambé cognac: sauté is fast and delicate and requires excellent timing and temperature control.

Frying means to cook in fat. You can shallow fry by cooking in fat or oil halfway submerged and turning or deep frying by fully submerging in hot fat. We cook this way mainly for texture. Coatings and batters make the difference here. Most chefs focus on starches and kid food when frying.

Grilling is cooking over a coal, open flame, or indirect dry heat source on a grate or crossbars. Most yachts and estates will have some form of grill. I do not count a grill pan as it is still a frying pan with some raised parts to make a cosmetic marking. Real grilling requires hot rising energy and a grate system that allows excess fat to fall away. In most domestic and superdomestic cases, the grill will be located outdoors. It will be rare that you find a nice DCS, Viking, or (if you are lucky) TEC grill indoors, but it does sometimes happen. A nice grill in your workstation makes your meal timing a lot easier as you will not have to walk sometimes hundreds of feet and go up and down stairs with your prep and finished products. Sometimes you will have electric grills in a designer kitchen, but most are just decorative and selling points at the time of purchase. They are weak and really just electric frying pans with holes. You will not get pro results with these. Some of your top-level yachts of 50–60 meters (164–198 feet) will have commercial electric grills, which are powerful and great to work with. When the yachts get bigger they tend to have a lot of government restrictions put on them about open flames or flare-up potential. At more than 200 feet, you will end up with large flattop griddles at a slight angle toward you with v-grooves carved in them to allow fat runoff. That creates the illusion, but it is not grilling.

Grilling has really taken of in the last decade as it allows for vegetables and proteins to be cooked with a crispy, flavorful crust while trapping the natural, succulent juices inside. The excess fat falls away, making for a concentrated natural flavor; it is also healthier than other methods of cooking.

I would like to add a few words about gas grills. Cast iron grates are king as cast iron takes a long time to heat up and subsequently a long time to cool down. It works as a heat bank. When you put your room-temperature proteins on the hot grates, the meat sears and sizzles and it does not suck all the heat out of the grates as it would with other metals, such as aluminum, which transfers heat very quickly and efficiently.

Stainless steel grates are the worst. They look great at the store (which is why they sometimes use them), but they are in that low-to-middle zone of heat transfer that tends to make everything stick to them. They also will never look good after their first use. Yes, the people who design your cooking systems actually design many items to be bought by confused amateurs, who hear words like *stainless steel* and *grill* and say, "This must be good." A stainless steel grill refers to the outside of the grill, and the best have cast iron grates. A crappy sheet steel grill with stainless steel grates is a way to use all those words in the same sentence in a showroom. BTU stands for British thermal units. They are a measure of power in the temperature world. Just like horsepower in the automotive world, think of BTUs as your grill's horsepower. You can never have too much horsepower—ever. There is a type of grill designed for superpremium proteins. It is called an infrared grill. There is a color in the light spectrum that allows for maximum transfer of heat, which in this temperature range gets blasted through time and space like a laser beam. It is that gorgeous orange glow coming off your charcoals and out of volcano lava. Infrared grills use a technology of a controlled burn through a ceramic brick with many tiny, evenly spaced holes to get the ceramic to light up as one large, evenly glowing source with no ash to block the light. There are no hot spots, and the heat is dry. Anything that drips on the glowing surface does not last long. It is said that the right infrared grill will take a 14-inch square ceramic brick with 1,400 holes and turn it into 1,400°F in four minutes, but it is actually hotter than that. This is what we use to cook a Lobel's 2-inch thick, prime, dry-aged shell steak fit for an actual king. The top steak-houses also use these. If you want one for your yacht or home, look at the TEC brands. There are also broilers on some yachts and estates. These cumbersome units, sometimes known as salamanders, heat from the top, and you put the proteins on the top heated grates below. They are designed to promote gravity basting, and once you master the technique they do fine. They are now considered old-fashioned, but if that's what you have to work with you better learn to use one. There are many books written on the subject of grilling, so buy them and experiment.

This is what I do with all-natural whole ingredients.

Get the grill super hot on one side and at a lower temperature on the other. Once again, start with a room-temperature protein or vegetable, season, and decide whether or not it needs an alternative fat source. Rib eyes, shell steaks, veal chop and such do not require anything but salt and pepper before hitting the grill. Leaner meats like tenderloins, pork and chicken definitely do. Olive oil usually does the trick. Fish, on the other hand, does not do as well with oil. Leaner fish stick more with oil and tend to pull apart on the grill. Fatty fish such as salmon will actually start to dump its own oily fat in a chain reaction if you start with oil. On fish use a liquid food release, such as PAM cooking spray. It is important to use only the original with no synthetic flavors to mingle with your superpremium proteins' flavor. Do not use the baking version, which contains flour. Coat the fish with an unbelievable amount of the spray, really soaking it, and then salt, pepper, and put on the grill. This technique works every time.

The first goal of grilling is to achieve a crust on the outside. Start with a clean grill. The grill needs to breathe so never cover the entire surface with what you are cooking, and no real chef in history has ever had aluminum foil on his grill. You need to leave some virgin space on the crossbars. Place the protein or vegetable on a 45° angle to the grates or crossbars. Once on the grill, the hardest thing to do is to leave it alone, but that's what good grilling is all about. There are no valid recipes to follow for the grill. This many minutes at this temperature does not work in real life. No two grills are exactly the same, and because of atmospheric conditions outdoors, most grills will never cook exactly the same way twice. Amateurs need that sense of temperature and time, but it is more important to learn what to look for. Read the protein: sight, sound, smell, and feel and taste all working at the same time. A great grill is hot and glowing; heat, gasses, and fats are violently making their way around the immediate atmosphere, and it should be hard to stand over for very long. When you start with room temperate proteins, things happen fast. You must imagine what is going on inside the protein and not look away for more than a few seconds at a time. All your senses should be alive and on edge. When there are hundreds to thousands of dollars on your hot grill you should have the same level of adrenalin pumping as a base jumper or the guy who tests Lamborghinis by drifting around corners with all four wheels smoking. This level of focus is awesome but not for the faint of heart. You must remain in control.

Flare-ups are not good; you do not want rogue flames jumping up to singe your proteins and cover them in carbon and soot. Flare-ups happen when there is too much fat on an item or you put oil on something already oily like lamb or salmon. This can work as a catalyst and start a fat waterfall onto the heat source. You can fight flare-ups with a spray bottle of water using a stream, not a mist, and then wipe the offending item with a paper towel to cut down on the fuel to the flare-up. When an item looks like it has that crust (Maillard reaction) starting to form, use long steel tongs to see if it is ready to lift. If it resists, the protein is trying to tell you that it is not ready to move yet. Give it a little more time. When it lifts easily, we move it to a virgin (clean) spot on the grill already at full temperature. The item is then ready to turn so that it is 45° to the crossbars (a 90° turn) in the other direction. Immediately use a grill brush to clean the spot from where the protein was just moved. This creates a new virgin spot on the grill that should take only a minute to be ready for use again. This 45° angle trick is really for aesthetic reasons only. It creates the nice, checkerboard cross marks and shows that you put some effort into it. Once we have created the cross marks and a nice sear, it is time to flip over and repeat the process. You can close the top of the grill or place a grill bell or cloche over the meat

FIGURE 5.8   PRIME CHATEAUBRIAND READY FOR PLATING.

on an indoor version to promote even cooking. This technique, once again, pays homage to the ultimate cooking method: roasting.

I have heard many chefs say to turn a good steak only once. However, since I am pontificating from a level of both privilege and experience (thank you, rich peoples' wallets), I will cook a 2-inch, dry-aged, prime supersteak on all sides as long as it is plumping evenly. Rules are for suckers. I follow laws: the laws of beef and physics. Cooking on all sides also works particularly well on beef tenderloins. As long as the inside is pink and even side to side, you cannot have too much golden crunch on the outside.

Vegetables and fragile proteins are cooked at a lower temp on the lower temp side of the grill. Once a protein is well seared, you can move it to the cooler side of the grill or even to a higher shelf or off the grill entirely to allow for carryover cooking take place. If the outside cools too much while the meat is holding for service, you can place it back on the grill just before serving for a minute to heat just the outside before plating. This is called flashing, and it can also be done in an oven, but only for a minute or two.

The trick with grilling is to pay attention, live in the moment and use common sense. If it does not seem right to any of your senses, it

FIGURE 5.9   FULL WHEEL OF REGGIANO PARMESAN CHEESE.

probably isn't. It should also not be boring. What a beautiful way to cook! Fish, chicken, and pork are cooked on the lower temp side of the grill but should not sit there lifeless, listen for the sizzle. Lifeless grills, like "well-behaved women, rarely make history."

Braising or wet roasting is the method of searing a piece of meat and then cooking it slowly in a liquid, submerged only halfway and in a covered pot in an oven. You turn the meat periodically, and theoretically you create a swirling moist and dry, controlled environment to render the meat hypertender and exquisitely flavorful. You see, the more a joint (a whole muscle) is worked when the animal is alive, the tougher it becomes but also the more flavor it will have. This long, moist, and dry method of cooking is the bastard offspring of both roasting and stewing. Osso bucco (braised veal shank) is the ultimate example of this. When I see a braised item on a menu, it is nearly impossible for me not to order it. The braising liquid is usually served with the dish. Stewing is where smaller pieces of meat are seared and then cooked fully submerged in liquid on top of the stove.

Steaming and boiling are not considered part of high cooking but usually involved in some of the sub methods of more complex cooking and fusion methods. We do still blanch in hot water, steam vegetables, and sear-steam greens in their own juices.

Molecular cooking can be fun but is considered more of a novelty by classically trained chefs as it usually detracts from the rancher's and farmer's original intentions.

If you look at any style of cooking, from peasant to king, it can always be broken down to the aforementioned cooking methods or a mixture of them. Just like a symphony, cooking can be broken down into many simpler parts requiring timing, skill, and restraint from all involved. The mastering of each of the previous methods coupled with a solid understanding of your ingredients from breeding to plate will give you a real chance at success in the elite private chef world.

CHAPTER 6

Menu Planning

Long ago in a culinary school far, far away, an instructor told me that if you want to open a restaurant you must first come up with a menu. From then on, every decision from décor to capacity to location would then be best made based on the merits of your menu. This is how successful food ventures begin. They say that in sailboat racing most victories are secured the night before when the crew are studying the rules and the weather patterns for the race at hand. Dinner parties and yacht charters are often a success due to the same amount of attention paid to details when planning them.

In private service, the menu is still everything, but it is often a case of the tail wagging the dog. When you open a restaurant, you are asking people to come on your turf to see what you have to offer. If you are good at what you do, 98 of 100 people will like or at least respect your offerings. The other two—well, either you or they are having a bad day. These are acceptable odds.

In private service, you have only one person to make happy, and it is all or nothing. Even if you work for a family, one person always holds your fate in their hands. If you please everyone else, and the boss is unhappy, they can still have you replaced. Again, the number-one rule in private service is, "Never forget who pays the bills." We are not crowd pleasers—unless that makes the boss happy.

The hallmark of a successful private chef is being a great detective. You have to find out what makes the boss and the people that they care about or respect happy—often without much direct contact. Many private staff will keep detailed records of the boss's likes and dislikes. This is a good start; however, the problem is that the boss's preferences may change over time, and many are based on the limitations of previous chefs that may be more skilled or less skilled than you. Some of your bosses may be royalty or sultans, and some may be crazy or drugged out: the bottom line is that you may not have a lot of personal time with them. The upper staff is your next line of defense, but bright-eyed, ambitious chefs may also be seen as a threat to longtime staff that do not want the status quo to change. They may feed you bad or out-of-date information on purpose until they approve of you or, worse, begin to plan your demise. The key to being a good detective is to glean your information from as many reliable sources as possible. Start with preference sheets if available, and then ask the other department heads. Next, seek out the boss's regular purveyors. If you know any past chefs and employees, you may interview them confidentially, but this can be risky. Perhaps your agent may have information from past employees' exit interviews. You don't just need good information; you need all information! Write down the things that you hear again and again. The truth will start to show itself. If you are doing a charter or a vacation rental, ask if you can speak with the main client for 5 minutes. I find that the tone of your voice and some basic questions and statements work wonders. You do not have to plan the menu on the phone, but you can get a feel for what you and they are in for. There are a few questions to always ask.

Are there any food allergies or special diets? Are there any special foods that they would really like to experience on this trip? Are there any strong likes or dislikes? Are there any special occasions or celebrations occurring during the trip? Will there be young children who do not eat with the parents? If they are traveling with staff from home, such as a nanny, when and which food does the nanny eat? You can sometimes inquire about their *foodliness*: do they require $100 a piece shell steaks, or would that be considered over the top for them? You can sometimes find things out without asking them directly; intuition is such a big part of becoming a great detective.

Charter brokers and high-end rental agents can be a good liaison and dial you in on budgets. Inexperienced ones may say, "Go all out!" but even without wasting an ounce your food and wine bill can easily rival or surpass the rental price in the hands of a serious gourmet chef. Sticker shock is never good for anyone at this level. I have had weeklong charters on superyachts with food bills of just under $200,000. Make sure that you are on the same page as the guests or the boss. I stand behind the theory that the worst thing that could happen is for someone to pay $1 million for a charter, pay the chef a small fortune and then say, "Is this all I get for my money?" Nobody wins when there is sticker shock; it puts a stain on the world-class experience you have just given them. It can also affect the gratuity if there is one or can even cost you your job.

Most charter brokers and high-end rental agents will have the guests fill out preference questionnaires. I used to teach a yacht and estate chef training course in Fort Lauderdale. Each time I got a new class, I would have them fill out one of these. We would pass them around and soon figured that even with such direct questions as, "Do you enjoy (check the boxes next to) Italian, French, Mexican?" other than allergies and celebrations we learned nothing about the complex nature of the way that a person who has no limitations on the cuisine that they can afford will in fact eat.

Once you do have good information on what your client eats, you now have to look at the limitations of the yacht or villa. If they want gorgeous shell steaks, do they have a suitable grill? If not, can one be brought in? You have to look at the storage capabilities and the staffing. Is the regular staff capable of advanced service requests? If not, can additional specialized talent be brought in? And, most of all, is the weather a factor? Are you planning barbecues on rainy days? Are you serving heavy meals and cream soups on 95°F days with 100% humidity?

Even if you are a genius at menu planning, the singular trait of the wealthy is that they have choices. If a friend of the boss pulls into the bay with an even larger yacht and invites your boss over for supper, all your menus for the week are now out of whack, and any defrosted proteins and preparations now have a different priority level than earlier in the day. Also know that the rich are sociable and more likely to invite extra guests on short or no notice. I find a weekly menu to be more of a guide. When you are on a yacht traveling waterways through micro-climates, some ports are missed, the vessel is rerouted, and sometimes the Mrs. spots the perfect bistro and just must have that experience. You can't take it personally; you just roll with it. Changing the menu is part of the puzzle you have to solve; it's why you make the big bucks.

While each boss is different, generally the industry standard for when the house or yacht is full is as follows:

- Breakfast: We set up a coffee station and a continental breakfast featuring a fruit platter, cold cereals, yogurt, and fresh baked pastries. We cook bacon and sausage that is three-fourths done and take individual orders as the guests like, *always* on their schedule. Most groups of wealthy people have different schedules and are used to getting their way, especially in a seven-star environment. We don't make restrictions as a good chef can poach an egg and belt out a single serving of Hollandaise while nailing cinnamon French toast and banana pancakes. The hard part is when a large group wants to eat breakfast together and all want something different. You just rise to the occasion.
- Lunch: Most people eat light at lunchtime—something along the lines of a green salad and a side of fish, a vegetable stir-fry over basmati rice, or a cioppino. It is usually one dish with a light dessert offering such as cookies or sorbet. However, it can be a long, drawn-out meal in courses if they are eating in proper European fashion.
- Hors d'oeuvres: This can be as simple as crudité and beggar's purses all the way to 5 pounds of cold, picked lobster meat from Legal Seafoods in Boston or a pile of stone crabs from Joe's in Miami. Hors d'oeuvres can be $50 a platter or $500 even if they are simple—and awesome.
- Dinner: This is generally a three-course offering with a starter followed by an entrée and a dessert. The entrée will generally be a protein, a starch, and a vegetable or two. Sauces are almost always on the side. Desserts are usually sumptuous and small, and there should always be seconds available.

Your menu should take all of these offerings into account. Most chefs do their menus the night before and leave it available to the butler or chief stewardess to view in the morning. Menus are usually brought to Mrs. or Mr. after breakfast and approved or returned with changes on them. Once locked in, you can move on with your day.

I like to use sheets to fill out that have the space for the meats, fish, vegetables, and starches marked out and then hang them in the kitchen or galley so that I can point and not have to answer so many questions. The service staff needs to know the menu as early as possible to set an appropriate table.

Because menus are only a guide until they approved that day, I find that it is better to have resources and varieties of food on hand to allow for as many changes as are needed and alternate choices for guests.

CHAPTER 7

Provisioning

So now that you have scored your new gig, have become a great detective, and have set a winning menu, it's time to go shopping. Provisioning is the most difficult part of being a private chef.

When I first joined the private chef world back in the 1990s, yachts were small, Michelin quality chefs in homes were rare, and the Food Network was just beginning to start the movement of gourmet groceries available to the masses. We would head down to the local supermarket and fill six carts with everything we needed for next two weeks for ten guests and five crew members. If we needed fine meats and fish, we hit up the local butcher and fish market. The captain would meet us at the checkout with the credit card, and the food was often beat up by the time it arrived at the boat. It was fun to see other chefs doing the same thing and barter over who got the last box of Cheerios. Local residents would interview us when they saw how much and how fast we were purchasing. The middle class would marvel at how we asked for the most expensive cold cuts instead of inquiring what was on sale. It was all we knew; it was all we had. It was South Florida, and whether we were cooking at the boss's house or heading over to the Bahamas with the boss's brand-spanking new 100-foot Broward motor yacht, we were living the dream. Whip up a few recipes from our last restaurant, and we were stars.

Things would change dramatically in a very short amount of time. The boats got bigger—a lot bigger. So did the homes. The whole yachting and private chef business quintupled before our eyes. No longer were the homes 7,500 square feet—that was now the size of the staff quarters and laundry. Entire new industries shot up around us to service our every need. Each chef was issued a credit card or at least held on to the boat card. We were given large sums of cash, $5,000 to $10,000 on average, and were told to use the card for everything we could and cash only when necessary. On joining a program, not only was it customary to hand over your passport as collateral but also to receive as much cash the chef felt that they needed as long as it did not surpass one month's salary. So a $4,000/month chef may have gotten that, and a $10,000/month chef could have up to that amount. You see, as the chef you were personally responsible for that amount of cash. If more was needed, the captain or estate manager would simply buy receipts. You would turn in your cash receipts and the leftover cash, and in return you got another big block of cash. Every time a captain gave you money, you would sign a piece of paper for it; this way the captain or estate manager didn't confuse the situation with something else and say that they gave you $7,500 instead of the $5,000 they really gave you. You would then buy back that signed receipt with your cash receipts and leftover cash, always making sure you tear up that slip on the spot so it does not end up back in the active pile.

I have seen captains on smaller boats stress over not wanting to take the laundry to a laundromat with big washers and dryers to get it all done in a few hours only to torture the stewardess who has to sit there with one tiny machine for 12 hours so as to not turn in a receipt. A chef will be able to smooth that over and take stress off the boat by picking up cash receipts from the floor at the local markets to replace the controversial ones. Trust in the private world is very important, so it is wise not to stockpile slips and to use them only with the captain or estate manager's permission. This avoids raising red flags at the accountant's office. Talking about a laundry receipt with the accountants and then having the accountant ask the boss why the onboard laundry isn't sufficient causes wasted time and tension about an asset designed to relieve stress, not cause it.

If the chef has a nice roll of, let's say, $5,000 in their pocket or in their purse and it disappears, they are still responsible for that money. It can happen to anyone and actually did to a girl around the turn of the millenium. Somewhere between leaving the marina and making purchases she became distracted and lost track of the envelope containing thousands of dollars. Regardless of how, where or when it happened she still had to face the consquences. It raises a lot of red flags, but most captains will take it out of their paychecks in installments over time. If the chef quits, their passport is held as collateral, and maybe they were stealing. You can sell them that passport for, let's say, $5,000. See how this all works? Of course, carrying $5,000 in one pocket or all in your purse is a risk, so if you must carry that much (and there are times you definitely will) spread it around your pockets and vehicle. It is simple risk management.

In the early 2000s professional provisioners started popping up to meet the needs of the new superyachts and monster estates. They started as small warehouses with walk-in coolers and a few desks, usually covered in catalogs, a phone, and a fax machine. You faxed over a list of foods for the chef, cleaning supplies for the chief steward and first mate and a list of parts from the engineer. They would show up at your boat by vans with stickers on the sides; most of what you ordered was substituted, and the billing was sketchy: $800 for cleaning supplies, $500 for steaks. Something wasn't right, and somebody was getting ripped off. We were busy; otherwise, we would have gotten this stuff ourselves. What do the boss and we have a lot of? Money. What do we not have a lot of? Time!

During this tumultuous period, boats were bought and sold at feverish paces, and often yacht owners had more than one yacht—one they were using and the other up for sale or for use as charter. We never had time to sit down and ask how much each item was. If we did, the provisioner would ask us to fax over an order, and they could maybe get you some prices after the season calmed down. That meant, "Don't waste my time." Consequently, yacht chefs don't shop; they buy!

I spend a lot of my free time in between jobs talking with and generally hanging out with my provisioners. They usually shower us with high-end free stuff, and they love to hear about what is really going on

FIGURE 7.1 NICELY ORGANIZED STOREROOM ON A YACHT.

out there in the marinas. I am very secretive about the details of what I do with most people in our industry, but I usually shoot pretty straight with the provisioners. They can help get you gigs, and they can save your ass when things get a little too real out there.

The big provisioners now have 40,000 square foot warehouses and a dozen rooms filled with cheerful problem solvers wearing Madonna phones (headsets). They throw sponsored parties with 5,000 to 10,000 people at boat shows that cost well into the hundreds of thousands of dollars. They spend a fortune to get our attention. If you are a department head, you are the point-of-purchase decision-maker. Many yachts spend millions of dollars a year operating with the average being 10% of the value of the yacht per year to operate. Fuel, dockage, insurance, salaries, food and beverage, cleaning supplies, and parts all add up quickly, especially when you are traveling internationally. A $50 million yacht will cost you $100 million at the end of 10 years. So yes, yachts spend that kind of money, but it does not all get paid by a home office. It is dished out by the department heads. If the chef spends $400,000 on food in a year, the chef is the one who decides where and how it is spent. We are lonely traveling types, working and playing hard on a global scale, and the provisioners know how to get our attention. Massive parties, hiring all the dancers from St. Martin's premier gentleman's club to come to our local crew bar to both serve us cocktails and put on a floor show. I still keep in touch some of those nice ladies. It is not uncommon for the provisioners to supply us with all the food and booze for a birthday party with 75 guests or maybe a bottle of a first-growth Bordeaux or dinner at the best restaurant within 100 miles. Maybe your whole crew has an open tab for the two weeks you are in port at the crew bar that they happen to own. Does it sound too good to be true? I'm not sorry—it really is that good.

Why do they spend so much money on us? A 200-foot yacht from a premier builder such as Feadship, Lürssen, or Abeking & Rasmussen can charter for $400,000 to $500,000 a week. They carry a crew of 17–20 who expect excellent food as part of their perks of the job. Most families charter for two weeks. So after the charter cost, food and beverage, dockage, gratuity, and private jet travel you are over the $1 million mark for a vacation. Your 12 guests (yes, only 12 guests on a charter or you have to reregister as a cruise ship and be inspected) expect nothing but the best possible cuisine. We do not mark up the food and beverage for the guests; we are paid to hunt and gather for them, and we do not have resale licenses. You must understand that no chef is better than their ingredients and you will not find any of these at the local markets: Beurre D'Isigny with the salt crystals, AA medium eggs, prime dry-aged meats, contraband cheeses, and poulet du Bresse are not coming to a town near you anytime soon. Your best restaurant in town may cost a million dollars or more to build, but it is not doing $1 million a week like we are. If you are, call me; I want to learn from you.

A private chef's primary job is to maintain the boss's standards of cuisine anywhere in the world. I have seen families have the meal of the century by full moonlight 20 miles out to sea off a Caribbean island on a sandbar that shows itself for only four hours a day. In these cases, it doesn't matter how much it costs to pull this off. It is a once-in-a-lifetime experience: $14,000 a head—no problem. That same dinner should cost a lot less in London or New York or you're getting fired.

How do provisioners justify their existence? They carry their own lines. Specific brands, of which they are an official dealer, are then marked up 18–30%, and you pay a somewhat normal price, much like a Sysco or a Monarch food distributor for restaurants. When you ask for specific items that they do not carry, they will send a shopper to find it

FIGURE 7.2   THE PROVISIONS ON A LARGE YACHT USUALLY COME BY THE TRUCKLOAD.

locally or track it down internationally, and that costs money. They also repack everything into their own boxes and cool-pack refrigerated items and dry-ice frozen goods. All this handling adds up. On top of that is the shipping; I once had a glatt kosher Passover charter that the food could not be stored with or travel with non-kosher foods. It took three 727 jet airliners to get it all down to the Caribbean, and I got handed a bill for $56,000 just for the planes. I had a team in the United States to ship it while I was in St. Martin with the island team to receive it at the Princess Juliana Airport. It was never out of our hands. You cannot believe how much it cost and how complicated it was. I had to have the provisioner rent and sanitize four refrigerator trucks to store everything while I amassed it. There are no kosher wholesalers because it goes against kosher laws. This means no discounts either, but there is nothing you can do about it. You may be able to throw a small kosher wedding in Manhattan for $100,000, but now move that same event to the top of Machu Picchu and see how the costs escalate.

Shipping is a huge deal; you cannot just FedEx everything. When food-stuffs enter different countries, they are subject to tariffs and inspections. I had an inexperienced provisioner (who is now out of business) miss a deadline with some very special heirloom tenderloins for the boss's birthday. They arrived too late, so we had to set sail to the Bahamas without them with the promise that they would be shipped overnight to the boat. The next day suppertime was coming and we had no tenderloins. We sent someone to customs and there they were, box opened, on the dock in the beautiful Bahamian sunshine. The beef was foaming, and $1,200 was wasted. I fired the provisioner—an upstart gourmet shop that did not know that you not only have to send them but also need an agent to receive them on the other side. Plus, they were the worst culprits of nondescript billing; you would put 50 small jars of gourmet products on the counter and get a total, handwritten amount of $2,156.97 for gourmet grocery. Many of the chefs in our industry, although very articulate, use English as a second language. They are not all used to American money, and most were not skilled accountants. This shop took full advantage of this. They were very generous with us; with all their small crimes adding up they could afford to, but at whose expense? The boss pays you well to look out for their best interests.

Free gifts for you or reward points for the chef that can be used only during certain periods are part of the provisioner's marketing budget and are thank yous for choosing that provisioner—most are samples from their distributors. These are the perks of being an international super chef. What you really need to avoid are kickbacks, which occur when the provisioner offers you direct cash for using them. Think of it more as a bribe: put $75,000 through my store this trip, and I will have an envelope with $5,000 in it for you when you get back. These kinds of things tend to happen when the chef is underpaid, abused, or taken advantage of by the owner and oftentimes when the chef is into drugs. We do take drug tests in the yachting business, but not nearly enough. Too many chefs slip through the cracks. I am not talking about smoking a doobie at a party; cocaine and opiates cause all the problems for these types.

The other way a provisioner makes money is working as a bank. Many yachts and estates have permanent accounts with these provisioners. Many captains and estate managers like to use them because it is easy and the billing is itemized and all in one statement. Also, we often do not receive funds for events or APA (advanced purchase allowance) until the last minute. The provisioners can carry hundreds of thousands of dollars in credit and debit for you. This is figured into the price, but remember: in our industry the real currency is not money but time.

If a highly paid chef has time to shop, they should. To go to a provisioner to get some items that they carry and you can't get other places or find cheaper is fine, but the rest of the hunting and gathering should be done by you. If you have a provisioner do all your grunt work and you have the time to do it, you are having the boss pay you to not do it and the boss also pays a provisioner's minion to do the job you are not doing. That is double billing and flat-out stealing. If you pay a friend out of your pocket to run errands for you (subletting) so you can do something personal, that is okay. Just don't bring it up or get caught. It is not stealing; it's just frowned upon. Rich people hate two things: waste and being taken advantage of. Don't take advantage of your good situation.

Not all provisioners are huge operations. Some are just people with a small office at a shipyard working seasonally. That does not mean you should discount them. I have been surprised at the lengths that some have gone to find nearly impossible to get items or pull in every family member to complete an order that looks way outside their business model. What do I look for in a provisioner? First is accountability: I want to meet the person in charge of my order. I want their personal cell phone number, and I let them know that if this all goes south I'm taking you with me. Second is trusting them when I meet them: I am not handing over $5,000 in cash as a deposit and giving them all our banking information on the dock after finding them in a local ad. They need to have a physical address that I can go to, even if it is just the office and they process offsite. Third, I need to talk to other chefs, the dock master, locals, or at least some local business owners to make sure that they are for real. If they are a huge operation and I am shipping internationally, do they have a satellite office in the location that I am shipping to? If things do not show up and you are going to ruin a million dollar cruise, can I call you at 4:00 a.m.? Will you answer me on a Sunday morning at 7:00 a.m.? If they say anything less than, "Yes, here is my personal number," try one who really wants your business and respects and understands your challenge.

Provisioners are really for programs that are large, remote, or cash-strapped but that have active accounts with them or with time restraints such as back-to-back charters. Most of my shopping consists of me in an SUV or minivan tearing up the countryside with collapsible plastic crates and a cooler in the back. I like to shop; I always have. I am in control, and I like solving problems that are not my own. I enjoy the interactions I have with my purveyors. I like when they offer me new things to try to show me new things that they believe I will like. If I have enough notice, I spread the shopping out over a few days: dry goods and cleaning supplies on day one; frozen goods and kitchen wares day two; vegetables and fresh proteins at the last minute. Try not to schedule inconvenient delivery times, such as when the crew is eating or during their personal hours. I do not like to overload the fridges and freezers too far in advance as there is no worse feeling then when you have tens of thousands of dollars worth of food stocked in them and they fail.

Accounting is a huge part of provisioning. You cannot exaggerate the importance of not losing receipts, cash, or the boss's credit card. I like to keep the receipts in one pocket and cash in another, not stuffed in with cell phones or a wallet. Shopping on your own is usually fast paced. As soon as I return I put all receipts in a gallon zip-top bag; this way, they do not get wet or blown away. I keep them in a cabinet in the galley/kitchen along with a few hundred dollars. That way, if I am offsite and there is an emergency and I cannot make it back, there is money for takeout to feed the staff or even in the case that I need to have someone walk up the street and pick up something before a store closes. Sometimes other departments that do not carry as much cash as the interior run out and need a 24-hour floater; no problem, but I make them sign a slip for it. Days tend to mush together when you are working at maximum load. The chef often becomes the banker and shopper for other departments. The chef will have the crew car more than any other crew member. A good chef will scan everything while driving; he not only will find out where the best butcher is and who has the freshest produce but will also note if they spot a marine supply shop or a laundromat. Part of the fun of being the chef is that you get to explore the communities you are working in and interact with the locals. I always carry a business card where one-half is a color photo of me in uniform and the other says CHEF and my name, then Yachts & Estates and my contact info. This simple card works as an ID and corroborates your story, and subsequent shift workers can identify you when you return to pick up an order placed earlier in the day. Also, we normally wear a T-shirt or polo shirt with the name of the boat and perhaps a line drawing of the vessel on it. You see, often the boat card has only the captain's name on it. When you have to get them to accept Captain John Smith's card and you are female, it pays to have as much evidence on you to organically back up your story. Always sign credit card slips with your own name, not the captain's. It is a legal contract, and this way you not only avoid forgery but also can tell who used the card that day if slips get mixed up at the purser or accountant levels.

When traveling, especially in remote island situations, businesses come and go, but American islands are going to have your big box stores and French islands will have your gourmet shops and the good wines. Use your stereotypes.

As always, your best resources will be your peers. Join the chef, private chef, and yacht chef groups on social media. Get the cards and contacts of your peers, even if you don't fully trust them. Sometimes they will want to show off their knowledge in a pinch. Plus, we all pass around gigs we cannot take ourselves. Facebook is a huge way to find out which of your peers are in the area or have knowledge of it.

Remember—when you are spending the big bucks in small, busy shops, don't wait in line. Demand that a manager take care of you and ask for case discounts and help loading the car (remember to tip appropriately). Learning about products and pricing on your own time or when you are not busy is fun. When it is go time, remember: don't shop; buy!

At the end of a trip or fiscal period, you will need to turn in all your receipts in report form. You will not get to all your receipts day to day, but every few days you should straighten them out and pile them in order. Each receipt should have the date and amount on the slip highlighted in a color code. I like green for cash, yellow for credit, and blue for on account. I also handwrite the date and total at the top and circle each one as when I am looking through a pile. It is easier to look only at the top of the slip, and this info is in a different location on each slip.

Some of these slips can be 6 feet long. The next step is to physically put them in chronological order: credit in one pile, cash in another, and "on account" separate. I add the cash top to bottom and then zero out and add it again bottom to top. They should be the same amount. To make this faster and avoid mistakes, remember that U.S. and European currency is metric and you do not need to use a decimal. Just make sure that you add cents to whole numbers (.oo) and add a decimal on to the final number. I write down each receipt in order on a separate piece of paper in columns and total them. By this point, you will have caught any mistakes by redundancy. This makes it easier for the purser to organize for the accountants. I sign and date the report and turn it in neatly in that zip-top bag along with any leftover cash. This usually takes about half an hour.

At this point you can get back to what you really enjoy: the cooking!

Chapter 8

Outfitting the Kitchen or Galley

It is customary to show a new chef some respect in the form of allowing them to make a few purchases to help make the galley their own. For a while, in the early to mid-2000s, yachts and estates were trading hands at a feverish pace. If the boss bought one and used it for six months to a year, it could be sold for more than they paid. Essentially, they were getting paid to use it. When a yacht or estate sells, the outgoing owner usually takes everything with them or distributes some of the equipment among their assets and staff. The new owner usually starts from scratch. Whoever is the first person using the kitchen gets to buy the equipment.

The problem is that chefs are expensive and are usually one of the last crew members to be hired. You used to find a hodgepodge of commercial restaurant supplies combined with items from a dollar store. You would find a 30-quart mixer and enough muffin tins to outfit your local bakery, a pre-WWII vegetable peeler left over from military surplus, and enough crappy, warped, paper-thin frying pans to make the boat lean to one side with no two being the same size. The captain's wife, an inexperienced hotel chef, or the head housekeeper, who cooked for the

FIGURE 8.1 FOOD COUNTER.

family before they decided to get an actual chef, usually committed the crimes. Often the original sinner was not comfortable with spending large sums of money or did not see a need for "fancy" items. In the case of the commercial mixer, a crossover hotel chef may have just gone with what worked for them at their last job—at a hotel with hundreds of guests. The trouble with not getting it right the first time is that most people feel guilty about replacing items that are not worn out yet and that throwing or giving them away is wasteful to the program. I have rarely stepped on a boat that had too little gear; in fact, usually you can't open the drawers because they are so packed with every one-purpose novelty from silicone garlic peelers to a set of six microplanes.

The private chef industry is high turnover by nature for many reasons. Each new chef comes in and adds a few personal favorites to the equipment pile, and over time it adds up. I will often come in and remove 60% of the gear in the galley and dump it right into the bilges or into a storage facility. A real chef who is working alone or with a small staff (my largest full-time galley staff had only six guys working under me) will learn to be ultra-efficient. We will not have dirty pots and single-use equipment piled up everywhere. We clean as we go and learn techniques to get the same results by using personal skills and reusing the same pots and bowls at different times throughout our

FIGURE 8.2   GALLEY ON A GIGAYACHT.

prep and service. Drop me in a strange kitchen anywhere in the world with a 10-inch French knife and someone is getting fed. I own some of the best equipment in the galaxy. Before 9/11, I used to travel with thousands of dollars worth of French copper cookware and diamond plate cases with knife sets that cost more than my first two cars combined. To this day, I only use a few. With a French or chef's knife, a paring knife, a bread knife, and a ventilated slicer—bang! Bob's your uncle. Every other knife I own is for show, and I own them only to put a scare into other chefs. I hate to say it, but if you show up with good gear and you look great in your uniform, all you have to do is not screw up and you will be a hero. This works everywhere in life. When you show up with no investment in your gear, people take note and it takes a lot of wasted time and energy to win them over. In the long run, it is cheaper to buy nice stuff that does not go out of style and take good care of it.

That being said, there are several different kinds of programs in the private chef world. A good rule of thumb is to take a good long look around at the room itself. Who sees this? Is this a highly decorated

FIGURE 8.3   GALLEY ON A MEGAYACHT.

country kitchen that the owners occasionally eat breakfast in? Is this a separate industrial zone deep in the staff's working realm? You can usually judge where to shop by looking at the floor and the counters. If the floor is a plastic composite and the counters are stainless steel, shop at a commercial restaurant supply. If the counters are granite and the floor is a plumb hardwood, a fancy kitchen supply shop at the mall is in order. Sometimes you will shop at a combination of both kinds of suppliers. You also have to assess how much the boss likes to visit your work area. What if you want nice stuff and are working out of the way of the boss's areas? This is where it pays to know what works well and how to buy it at the right prices. Before shopping, you need to make a list of gear that works well for the amount of use and for the volumes that most yachts and estates are using. Let's use 12 guests as the gold standard. If you have more than that, you are commercial. If you carry less, you can adjust.

The average galley or kitchen requires a 12-cup food processor, a 6-quart stand mixer, a stainless steel blender, an immersion blender, a four-slice high-grade toaster, a forged stainless steel knife set, and a magnetic knife bar. You will need a large cutting board, but which one? Thick, eastern, hard-rock maple has a great feel but requires oiling and tends to warp when wet on more than two sides. Plastic has a crappy feel and leaves grooves. Bamboo is getting popular because it doesn't warp and is close in feel to wood. And now there is even a tan, rubbery board that can go in the dishwasher and come out bendy and hardens up when it cools on the counter. It is fun but has a miniature golf feel. Most of us have a large board that lives on the counter as the center-piece of our workstation and thin plastic cutting sheets that we put on top of our boards when we are working with raw fish or chicken. Those go right in the dishwasher for cleaning and sanitizing.

Pots and pans are where the big money is spent. This is where you have to think to yourself: how long am I going to be the chef here, and is what I am getting going to be dead in a year or two? If so, did I pay too much for it? I have owned a few Mercedes Benzes in my time. I like them because as long as you maintain them properly, 20 years later they still look and drive the same. They never truly go out of style, and you get a consistent experience throughout the life of the car. If you buy a lesser make of car, not only does it go out of style but also the quality of experience is always in a slow downward spiral. Going el cheapo here is rarely the answer. Going for French copper is sel-dom the answer either. Nothing can touch the quality of copper; it gets a 94 on the heat index. Put 1,000°F on one side, and the other will be 940°F. Thermally, the next best metal is aluminum, which gets a 57! With copper, if you have a boiling pot and slide it off the burner the boiling stops: back on, it boils; half-on, half-boil. Copper is the sports car of cookware. If you are that good of a chef and have that kind of focus, this is as close to drifting a Ferrari in a kitchen as you are going to come. Copper is profoundly expensive and heavy and requires not only a hand cleaning but also a chemical treatment and polish. It can never go in the dishwasher. Copper can take heat and chemicals but not at the same time. Copper is a permanent investment; when they found the *Titanic* the copper cookware was spread out on the sand next to it and ready to use that day.

Aluminum pans cook well and evenly. They are cheap to buy and the darlings of restaurants with big gas stoves that don't care what shape they are. Aluminum warps easily and can react with some foods that can cause a toxic reaction over time. Stainless steel is a lasting invest-ment and can go in the dishwasher, but it is a poor conductor of heat. It builds up heat slowly and cools down slowly and has a reputation for hot spots. It is nonreactive and great for boiling. Today we have some great compromises: alloys and hybrids. The pan that seems to work in most situations is the All-Clad LTD line, which features an anodized aluminum body lined with a thin tough layer of 18/10 stainless steel. It solves so many problems found with single metal pans. The aluminum content makes the pan cook quickly and evenly and is lightweight, and the stainless steel interior is indestructible. Thus, the caveman engineer or the gardener won't destroy them if they volunteer to help with the dishes. The stainless steel is too thin to build up hot spots yet strong enough to keep the aluminum from warping. Read the manual before tossing any pan in the dishwasher. You can find deals on these sets online.

You will need two 12-inch, two 10-inch, and two 8-inch sauté pans. The only nonstick pans you should have onboard are two omelet pans. I also highly recommend a good-sized cast iron skillet for searing. Cast iron is great because it is a terrible conductor of heat. It is like a bank for heat and energy. Cast iron takes forever to heat up and therefore forever to cool down. The cast iron will sear the meat without the meat draining all the energy from the pan.

You will need two large stockpots, two 8-quart pots, four 3-quart pots, two 1.5-quart pots, and a roasting pan. Be sure to have lids for all the pots.

FIGURE 8.4   GALLEY ON A 100-FOOT YACHT.

FIGURE 8.5   FOOD TAYLOR'S KITCHEN.

You will need a series of stainless steel prep bowls. A few should be large enough to mix a salad for 12 and many small enough for your mise en place (i.e., standard prep). For baking, have two muffin tins (not 20), maybe one for mini muffins and one full size; eight half-sheet pans and four racks that fit in them; two 9-inch springform pans; and one pie tin. Also include three Silpat nonstick baking mats, a rolling pin, four silicon spatulas, and two 18-inch pastry bags with large star tips.

Pick your own hand tools (whisks and spatulas), but real chefs must have great tongs—steel, rigid, and locking (Oxo makes a nice set), and in several lengths.

If I start with a bare kitchen and this list arrives, there are no excuses. Everything else is gravy. I am an old Yankee, and I know two things: quality and a bargain. I know what I want, I won't settle for less, and I won't pay top dollar for anything unless there is a time constraint. In the private service world, time trumps money every time. You see, if you regularly circulate HomeGoods, Marshalls, and T.J. Maxx, you will find these items scattered around. You can't count on them for a whole outfitting, but you can build up your collection if you have the time. Even if you buy these stores out, they will have a few more pieces the next time—and for pennies on the dollar. Most programs are not that money conscious, but if you are responsible and make it a game you can have anything you want with no major hassle from the boss.

If you have established yourself in the hierarchy or you are in the right place at the right time, the boss may entrust you to help refit the galley or kitchen. This can be a blessing or a curse to all who follow you. If you do not have the experience; do not participate. I have seen $50 million yachts with an inexperienced home cook dictate disasters. So much money was spent that it was too expensive to correct. On these boats, for the next 10–20 years they turned over chefs and frustrated the crew and owners. Using a yacht as an example here—because most estates use designers to do their kitchens and the equipment is not quite so built-in—a galley is the heartbeat of the boat and is a busy place. We not only have to cook breakfast, lunch, hors d'oeuvres, and dinner for the guests but also have to cook lunch and dinner for large crews. The boat moves, the galley rocks, and we do not have the ceiling height that so many estates enjoy. When people are on a yacht, they are most likely on vacation and active, and the guests may be from different time zones and may have children accompanying them that require meals at separate times. Yachts have a positive pressure-controlled climate, and any extra heat causes strain on the system. We experience long hours in these galleys because we do not have relief chefs and room for extra kitchen help. When an inexperienced chef comes in and says, "I saw a show on French hotels and they all use flattops; therefore, I shall get a flattop!" it can be a real disaster. Flattops are great in French hotels because they have high ceilings for the heat to dissipate and are high enough to wear a chef's toque that makes the heat go right past your head. They are great when you have meals à la carte all day long and need a variety of pots to fit in a small space. They are great when you can use all the natural gas you want or the city's electrical grid. On a yacht, people have requests all throughout the day at odd hours. Flattops take an hour to come to temperature, and there is little control over the temperature once there. They use a ton of electricity, which the boat has to make and usually has in short supply, and you have to keep them on overnight in case of an early egg order. They throw off a huge amount of residual heat similar to that of a radiator and that heat collects at head level, which is what makes chefs tired and angry. The air conditioning has to fight this 24 hours a day, and all that heat can be bad for the exhaust system. You cannot put fiddles on a flattop; therefore, you cannot use the flattop at sea.

Fiddles are a crossbar system that lock into holes around the stovetop and can be adjusted to lock pots into place when the weather gets rough at sea. This stops a hot pot from being launched across the galley like a death missile when we hit a big wave. Yet some bartender-turned-private-cook put one of these on a $50 million boat because no one said he could not. A better choice would have been two high-power, normal glass top burners and six induction burners. The two normal burners can be used with any pan. The induction works with many types of pots and pans, heats instantaneously, throws off no additional heat, and cools instantly. You also have maximum control over the temperature while using minimal electricity.

FIGURE 8.6 MY KNIVES. I MOSTLY USE THE RECTANGULAR ONE;
THE REST ARE MAINLY FOR SHOW.

FIGURE 8.7 GALLEY ON A 140-FOOT YACHT.

People eat healthier today, and a steam cabinet may be a better idea than a fryolator. Unfortunately, I see a trend with inexperienced chefs removing good equipment such as steam ovens because they do not know what they are or how to use them. An indoor, high-powered electric steak grill is a must if you have the right hood system, and two

FIGURE 8.8 GALLEY ON A SUPERYACHT.

identical convection ovens under your identical burners are enough to make any chef swoon.

It is so important to choose the right chef first. So many are on the way up or a bargain, but simple mistakes can really take their toll down the road. How many chefs will blow up at a stewardess just because they are going nuts in an unnecessarily hot galley that is getting to them after 20 days straight at 18 hours a day? Captains, other chefs, and engineers need to look at the proposals by the chef and then do their homework on how efficient these choices really are.

CHAPTER 9

Breakfast

They say that where and the way that you wake up can shape the rest of your day. This is your chance to really set the boss up for a great day. I once worked for a family where Mrs. was friends with music star Sting's wife, Trudie Styler. She said that Trudie's chef (who was female) would meet with her over coffee in the morning in one of the private areas of the house. She alleged that the chef could read Trudie's mood during that time and plan a menu that would "balance her out." This is private chefing at the highest level and one reason that there are so many successful female private chefs. I am a very capable chef who can handle almost anything, but I do not read my boss's emotions while relaxing with them. No one wants a giant man in uniform hanging around their wife's bedroom area in the morning. Female chefs have a real advantage in that department. If you can affect your boss or their family's mood by their diet, the odds of success are stacked in your favor.

Nowhere are rudimentary cooking skills more valuable than at breakfast. In my 15 years of feeding the ultrarich and ultrafamous, I find that great fusions of ingredients and special options are the exception during breakfast. The boss will usually prefer someone who really knows how to deftly poach an egg to the wizardry of the molecular pancake. Every client is different, and they will let you know if they want the same thing most days or some variety. Because you are dealing with a mix of people with different lifestyles and coming from different time zones, it is your job to cater to them rather than having them adjust to your schedule. If you have a full house on land or at sea, there are definitely some industry standards.

A basic coffee service, featuring coffee, decaf coffee, a tea box, hot water, and accouterments, is usually set up early by the service staff and refreshed throughout the morning. Just before the group starts to rise, the service staff will set a continental breakfast buffet. This will usually consist of an assortment of dry cereals, yogurt, a fresh fruit platter, and granola. Europeans will appreciate some cold cuts and cheeses. The most important part of this continental buffet is fresh-baked breads and pastries made in-house. There is a huge difference between breads and croissants that were baked up the street, handled several times, and then transported in bags only to be pumped with air like a bellows in transit that is losing freshness with every step. If you have never had the pleasure of waking up in a house filled with the aromas of fresh-baking baguettes, croissants, panna chocolata, and quick breads, I hope this comes to pass for you. It is truly one of life's richest treasures. It makes you feel like there is a better place than your warm comfortable bed within reach. No matter what problems you struggle with in this world, most can be overcome by this—and bacon.

At the sight of the first guest, we squeeze fresh oranges. Another of life's great experiences is real fresh-squeezed orange juice, and I'm betting that most of you have never had this done properly. There are special juicing oranges that are ugly, and we handle them roughly to promote juice separation. True fresh-squeezed orange juice, like fine gourmet coffee and scones, are all at their best only for about 30 minutes. You can make additional batches as needed; the servers usually handle this, but if you have the time you can help.

As a chef, you will be responsible for the fruit platter: no rinds, no peels, and it should all be in bite-sized pieces or easily broken down for eating with a fork. As a rule, the wealthy do not eat with their fingers. There should be nothing they have to discard, leave on the plate, or have to work at to eat. We do the work for them; that's why they pay us. You may garnish the whole platter as you see fit, but you can never

FIGURE 9.1 TYPICAL MORNING BREAKFAST PLATTER.

go wrong with fresh flowers. They remind us that food is beautiful and from nature; they look and smell nice, and some may even be eaten. I would not get hung up on using only edible flowers as a cheap bouquet from a local market coupled with a pair of scissors can go a long way. If the boss has impulsive children though, you can go with edible flowers exclusively.

The chef will also be responsible for the charcuterie and cheese platter if there is going to be one and of course for the breads and pastries. Some of you will think to yourself, "I am a chef, not a baker." However, I'm here to tell you if you can't bake, you are not a chef, at least in the eyes of your peers, so get to work on that. If you are not a real chef yet, it is not a problem. Many of us do not have the right equipment or space to do such things correctly. What I suggest is that you find a reputable baker that does everything from scratch and have them make all your bread and pastry dough for you. Have the baker freeze the raw and unproofed doughs immediately. Before you go to bed, you pull out your frozen doughs for the next day and put them in your kitchen/galley on your parchment paper or Silpat-lined half-sheet pans. Leave them out overnight so they can thaw and rise slowly, giving them more character. In the morning they will be proofed and ready to bake. Your pastries get a quick brush of room temperature egg wash (50% eggs/50% milk),

FIGURE 9.2   MORNING FRUIT PLATTER, IN PRIVATE SERVICE. GARNISH NOT NEEDED
EVERY DAY. THE QUALITY OF THE FOOD IS ENOUGH.

remembering that it is important that you do not use cold egg wash as cold makes things condense and shrink and would make your pastries fall. I like to do the pastries first as they take less time and get to cool a little before we place them lined up on a platter or in a bread basket. Also, this way no one will have to wait for them. They are handled only once by you and once by the guest. If someone orders toast, I use the bread left from the day before, which works fine. Once the pastries are cooling, the current day's breads go in the oven.

When the bread is done, it is time for the hot meats to go into the oven. Bacon and sausage are only as good as the animals that they came from, and the best bacons can be over $10/pound. If you do not know your boss's preference, thick, center-cut applewood-smoked bacon is the choice to make. I like to take a skewer, pinch the end of the bacon to the end of the skewer, and spin it around like a Maypole. I then wrap my hand around bacon and give it a little squeeze and then slide the skewer out while placing the bacon on a rack in a half-sheet pan. We call this pigtail bacon, and when baked it gives you a better yield. It is crunchy on the outside and juicy on the inside and cooks evenly. Pigtail bacon looks great on the plate and is easy to portion. Sausages will also retain more juice when done in the oven. Most chefs run a 400°F convection oven during all meal times. The convection oven is a regular oven with a fan in it. It helps to distribute the heat more evenly in an oven and to vanquish hot spots. It does, however, add about 25°F to the way the food sees the temperature. This is due to the efficiency created by moving the hot air; think of it as a kind of reverse wind chill factor. Heat still rises, so whatever is on top is going to brown more. We do not cook the meats all the way through. They are done halfway and then kept aside to be finished when ordered.

There are three types of hot food service at breakfast. The least used method is buffet. If you have a full house that has a scheduled activity or are the chef to a sailboat race team and cooking for them on the mother ship, this is a way to take the liability off the cook and put it on the individuals.

The second least used method is the group breakfast with individual orders. The private service world is not run on the word *no*. If it can be done, it will be done. That is what separates us from hotels and restaurants. If the boss wants to eat with all his guests and for each of them to have their own order, we will do it. The trick is that we have to get the service staff to go around to each guest and get the orders in advance before they sit down at the table. Pancakes and waffles are cooked ahead of time and put overlapping on a sheet pan. Eggs are poached, shocked in ice water, and then kept on the side on paper towels. Toasts (raw breads or crumpets) and meats (already precooked halfway) are put together on racks in half-sheet pans then put in the oven together. Toast in the oven is better than toaster toast, it is lighter and fluffier, and the room temp French butter seems to be very at home in the crust. Real maple syrup is put in the microwave in a porcelain or glass pitcher (this does not take long and can get really hot, so be careful!) and later transferred to ramekins or bullets. Take a small stainless steel bowl and put it over the poaching liquid to make Hollandaise sauce; remember that a breakfast Hollandaise is just egg yolk, fresh lemon juice, salt, pepper, and the clarified butter or ghee, which you should always have ready at breakfast. Plates should be hot; just put them in the oven for 30 seconds. Pancakes and waffles can be flashed on the half-sheet pan you stored them on. Fried eggs are to be done last when everything is on the plate so the plate does not have time to overcook them. A fresh orchid and a few berries and your servers (who should be standing before you ready) can whisk them straight away to the guests. Wow— fast and demanding on a chef, but when it's over it's over.

The most common way for a group to eat breakfast is for them to come up individually and have coffee and something from the continental breakfast bar, read the paper, and when they are ready (maybe after a run or a swim), order a hot à la carte breakfast. The main seating area will be fully set, minus plates. As each guest finishes their meal, the plates and setting are removed.

Want to know how to kill at breakfast? Here are some simple tips and recipes that will take you over the top without resorting to restaurant menu gymnastics.

1. Freshly squeeze orange juice when you see the first guest on the way to the service area, as it is only at its best for 30 minutes.
2. Offer French presses for coffee, but remember: the coffee must be ground more coarsely than regular coffee, the water must be a healthy simmer but not a boil, and you stir then plunge twice. Plunging is done at the table. If you use finely ground coffee, it may clog the screen and shoot up all over the guest; at the very least grounds will make it into the cup.
3. All eggs must be AA medium only and at room temperature. Put all the eggs you'll need for the entire day out the night before. AA eggs have a stronger, better-defined yolk. Medium eggs have a better yolk-to-white ratio and tend not to run or break apart under their weight. Room-temperature eggs cook faster and more evenly than chilled eggs. Chilled eggs condense, and room-temperature eggs are not unsafe as long as they are kept in the shell to the last minute and used promptly.
4. On special occasions, have an omelet station. You can make an omelet every 30 seconds if the eggs are at room temperature. Whisk the eggs and put them into a hot pan painted with clarified butter while shaking the pan back and forth. Never stop sliding the pan back and forth. The omelet will form and bunch up like a cigar. Slide the omelet onto a hot plate and add

a hot topping. You will impress your guests when you knock out 10 omelets in 5 minutes. Preheat your toppings. Crack and whisk all your room temperature eggs right as the line is beginning to form.

5. Spend some time on your fruit plate and make it look nice. Every piece should be able to be eaten with a fork and add lots of flowers for garnish.

6. Learn to poach perfect eggs. They can be done in advance, shocked, and then reheated in the poaching liquid later. The eggs are the most important part—again, AA medium, room-temperature eggs. The poaching liquid is just simmering salted water with a splash of white vinegar. Crack the egg into a coffee cup, lower it into the poaching liquid allowing the beginnings of a skin to form, and then dump the egg right into the poaching liquid. When almost done, remove with a slotted spoon, and either use or shock in ice water and store on dry paper towels.

7. You get what you pay for with bacon, sausage, and ham.

8. If service will be drawn out or done outdoors, set up a nice Plexiglas pan with sides and a drainpipe. Fill the pan with crushed ice and place all the yogurts, milk products, and fruit platter in it; you can put a tablecloth over the ice or just garnish the ice. You can obtain these pans at restaurant supply shops as they are commonly used for ice sculptures.

9. At breakfast, condiments are king: fine French butters, Swiss jams, and nothing but the finest, real maple syrup (which must be served hot!). When the foods are simple, you better make sure that every component is the best.

10. In a pinch, if your pastries fail you can take a can of buttermilk biscuits, cut the dough in half, and carefully deep-fry them. When you take them out of the oil, you can roll them in sugar and they make convincing beignets. People go crazy over them.

11. Sometimes champagne or Prosecco can really make the morning. Have the glasses on the table and the bottles chilled in sight. If it's already there, they will want it.

12. Play some soothing music, something relaxing that your boss already likes. Keep the volume low.

13. Smile. Find a way to enjoy the morning. I like to come in early and enjoy my coffee before I get started. People tend to mirror back the attitude that you are presenting. If you are happy with the morning, maybe that feeling will be contagious.

14. Never skimp on coffee. Pay whatever it costs to get the best in the style that the boss wants. Don't skimp on staff coffee within reason; it is not only a small luxury but also one that will be appreciated and a motivating factor in the way your colleagues start their work.

15. Make friends with your breakfast work partner (server). Make sure that they know that you will secretly make them anything that they want for breakfast. There is nothing better than arriving in the galley/kitchen and your coffee is on your cutting board they way you like it and your oven is already turned on. People get married for less. Keep breakfast hours mellow and hassle-free.

16. Keep a bowl of room-temperature French or cultured butter available at mealtimes. Use a pastry brush to paint toasts and the top of pancakes with it. White toast and top-class pancakes are my two favorite foods, and it is all the fault of this type of butter. Noncultured or so-called American butters contain too much water and usually lack character (foreigners, hold your laughter). If you cannot eat a spoonful of the butter plain, do not use it.

17. Communication must be efficient the morning as private service workers rarely get a full night sleep. I do not want to have my service partner running through the door barking out complicated orders over my morning James Taylor music at a speed I cannot understand. Something is eventually going to be forgotten, and a guest or the boss is going to be let down. I preprint sheets that simply have the guest's name, eggs, meats, starch, and syruped items with a line after each. You can put 8 or 10 breakfast forms on a page. The server will fill it out and simply say, "Order in." Guests' names are important as they may move around and the server may get called away as the fresh meal comes up. If a server is not around, you bring the food; most people do not mind seeing the chef for a moment here or there. Smile. Whoever brings the food out needs to put a line through

the order on the sheet as not to confuse it and have someone else bring the same meal out to guests. Sometimes a day-old sheet pops up and we look incompetent when we show up with yesterday's meal—again. Your service partner should use a small notebook and write down the orders. I do not know why people think that part of good service is having a photographic memory, as if everyone with a mind like an elephant immediately goes into food service. I feel better when I know there is a paper trail, and I never want to see a server come to me a second time to have me repeat an order.

THE ROYAL FAMILY'S SCONE RECIPE

One of the great privileges in working with and for the super wealthy and ultrafamous is the accessibility of top class information. Whether it is stock tips, real estate trends down to the month, or just a highly prized recipe, we overhear it all. Years ago, I worked for a family in Palm Beach that was friendly with a royal family (not going to say which one, but one that values scones). When royalty travel, they bring along what is called a valet. This is a personal travel butler who works behind the scenes to ensure that their boss's needs are met in your boss's home. They also oversee the standards and make sure that their boss is in no danger from poor food and beverage handling. They also bring along the royal's personal recipes for their must-haves.

The royal scones are not like the large, tough scones from a common bakery: they are light and hot, and they have a distinct bite that comes just after the French butter and Swiss raspberry jam slide over your tongue and the scone itself kicks in. They look like more of a buttermilk biscuit, which has been risen from the inside by way of lady angels breathing into them and flapping of cherub's wings. They are only heaven-sent for about 30 minutes.

They are not, fortunately, hard to make. Without further adieu, I hereby share the *Royal Scone Recipe* that I absconded with in my head so many years ago and have been using to enhance the lives of my clients for so long:

Combine in a food processor:

> 2 cups cake flour
> 5 teaspoons baking powder
> 1 teaspoon caster sugar (regular will do in a pinch)
> A pinch kosher salt

Run the food processor to mix the dry ingredients. Then stop and add:

> 4 oz. cold French butter, cut into small cubes

Hit the pulse button 5 times for 1 second each time. The butter should still be visible in smaller portions mixed throughout.

Mix together thoroughly:

> 1 AA medium egg, room temperature
> 2 oz. buttermilk, room temperature

Add the milk/egg mixture to the food processor and turn on for 10 seconds or until the dough start to roll into a ball. *Do not over-mix.* There can be a few small dry patches; the moisture will seep through the dough.

Place the dough on a flour-dusted counter or cutting board and roll out 1 inch thick. Cut out the shape of scone you like; I like a 2-inch circle for breakfast and a 3-inch circle if I am using these for strawberry shortcake.

Bake in a preheated 425°F oven without convection. They should be lightly golden on the top; this usually takes around 20 minutes.

Serve with French Buerre d'Isigny (with the salt crystals) and Hero brand Swiss raspberry jam. They are at their best between 10 and 30 minutes out of the oven. Try with a French press coffee and a glass of real, freshly squeezed orange juice, and you will have peaked for the day or even month.

CHAPTER 10

Lunch

Lunches are lighter in the private service world, even with Europeans, who normally eat their big meals in the afternoon. When they are having full breakfasts and loaded schedules, they seem to be wealthy first and European second. Every boss is going to be different, but I have found that most eat one or two courses at lunch as opposed to the standard three or four at dinner.

Lunches can be as simple as a gorgeous salad platter and a whole side of grilled fish filet (boneless, skinless, and drenched in fresh lemon juice) or as complex as a beach barbecue with 15 or more items. The trick is to keep a nice balance between hot and cold foods as well as a variety of textures and acidity levels as to aid with digestion. Those who have had a full breakfast and are about to embark on a busy day require it. This is the meal to use lots of ingredients, textures, and flavors: from a variety of salads with fine vinaigrettes to stir-fries featuring Indonesian soy syrups, petite vegetables, and a base of oil, ginger, garlic, and scallion. Michelin star cooking focuses on a few simple top-class ingredients prepared with a ton of respect and a profound level of attention to detail. Save this energy for dinner. Most people do not want to be blown away at every meal; they want to be fed gorgeous foods that make them feel good after eating them. Save the high-focus plates for dinner to avoid burning you and the boss out.

It is important to have a good mise en place at lunchtime as most foods are cooked and prepared à la minute (at the last second) to get the full effect of freshness and to not have salads sag or flavors bleed.

Lunch is the meal that takes place in the middle of the day's activities, so timing is very important. Meals should come out fast and on time. If there is a dessert, it is usually something light and fast, maybe a cookie platter or a sorbet. I find that cutting and toasting slices of a panettone in the oven and serving them with small scoops of vanilla gelato can be a real winner and can save you in a flash if you did not plan for a dessert or the boss changes their mind and now wants one.

If you are on a yacht or your boss lives on the water, you can always break up the monotony by presenting a beach barbecue. This can be confusing for those without a lot of catering experience. One thing to keep in mind is that coolers are like thermoses; they keep cold things cold and hot things hot. You can par or fully cook much of the meal and put your hot items in a standard cooler (a 160-quart Igloo type is the workhorse of our industry). You can boil or bake ribs or grill tenderloins, steaks, or chicken back at the yacht or estate and then bring them to the site. You can finish or flash them onsite on the grill. Your salads should come in gallon zip-top freezer bags in another cooler with the trays and garnish brought separately. Barbecues can fail for two reasons: poor planning and communication and lack of delegation ahead of time; and not thinking big enough when it comes to equipment. Yachts and mansions are not cheap; they never were and never will be. The boss does not buy these high-ticket items to make or spend money; the boss buys them to have a quality experience and to live life at a higher level. If you are a private chef, captain, or estate manager, part of your job is to know how the catering works and to make sure that cheap purchases do not ruin the boss's experience and waste their precious time. I have been on yachts of staggering proportions and have shown up to grill at the beach only to find that I have a small Weber grill and a bag of Kingsford or a $59 gas grill designed and built for one-time use for a family of four. I am not usually cooking for four; sometimes it's 64, and they want choices! Three hours later, half of the food was cold and I was out of Kingsford. Buck up and do your research. TEC makes a suitcase-sized (although heavy) infrared grill with 14-inch square grates that can get to 1400°F in about four minutes

FIGURE 10.1   GIGAYACHT BEACH BARBECUE IN ST. BARTHS.

with no hot spots. A pair of these grills can handle anything. They can run off large grill tanks or disposable cans. There is an optional stand, but because heat rises you can put them on top of almost anything. They are stainless steel and built to last. They are not $59, but if you can taste the savings you have already lost.

I cannot tell you how many times I go to do a beach barbecue and find that the boat has six coolers, ranging in capacity from a 12-pack of beer to big enough to put half of the fish caught on a small fishing trip. People, if you want to win in this game, think big—not huge, but big. 160-quart coolers hold a lot and can work as benches for sitting or working. They are not eyesores, and if you have a smaller yacht they can live in the tenders when not in use. Beer, wine, water, hot and cold foods, and later the dirty dishes after the party all fit nicely in these.

Provided you have the storage, include sturdy tents with roll-down screen sides, lighting, a decent sound system, and a generator; do what you have to do to make it awesome and memorable. The trick is to get everyone involved and engaged days in advance. I once worked with

FIGURE 10.2   FRESH LOBSTER ON THE LUNCH MENU IN ST. CROIX.

a crew where all deck personnel had their own shears and machetes. Once we docked or anchored somewhere, the first mate would arrange a quick hunting trip for the best palm fronds and flowers from the area. They fully decorated the flybridge and any beach tents.

Air mattresses with quality sheeting, driftwood platters next to them in the sand for drinks, big beach umbrellas, and a gas-powered blender can really add to the experience. If the yacht has a spa, bring the masseuse. Hire local musicians to play, or if the crew has talents (and they always do) have them entertain. Anyone can feed a boss at home or on the boat, but when you do a Michelin star quality meal at the beach, in a one-of-a-kind location at a small village built only for their pleasure and on only one occasion, you have now created an unforgettable experience. Keep in mind that this is one of the only times that the boss gets to see their multimillion dollar yacht from a distance, on its own and in its element. Imagine how that must feel to them.

Just because you are cooking at the beach does not mean that the food has to be less than spectacular. The chef comes to the beach in a chef coat even if a swimsuit or shorts are on as well. Once you are there, you are on display; be prepared, and do your stuff!

FIGURE 10.3   CHEF NEAL SUPERVISING A BEACH BARBECUE IN ST. BARTHS
WITH OUR YACHT IN THE BACKGROUND.

In some locations, bugs can be an issue, so think ahead and send a scout. The chef and chief stewardess should have a list of who does what for every detail. These lists should be passed around to the crew well in advance. If the barbecue is postponed to another day, the plan should not change.

I know we are talking about lunch, but some of the best barbecues are at night. Nothing beats a moonlit beach, a fine meal, great bottles of wine, a decorated tent, and a distinct lack of bugs. Day or night, if you are going to use a generator, use long power chords to keep it distant. Play music, and do what you have to do to direct the sound in another direction.

Once lunch is done and the chef has cleaned up, it is usually off to the market, a little personal time, or a good catnap. Do a study of naps on the Internet; there is definitely a science to them.

CHAPTER 11

Hors d'Oeuvres

It would be difficult to write about cooking for billionaires without bringing up the hors d'oeuvres or amuse-bouches. There are just so many occasions where the boss is going to have guests drop in or invite some friends over for cocktail hour. Most times, when you have a group of four or more there will be a proper cocktail time lasting one to two hours preceding dinner.

The term *hors d'oeuvres* translates literally to "apart from the work" (or the main work). It is a bite-sized portion of food made with the goal to awaken the palate and to take the edge off hunger. It also aids with the cocktail part of the cocktail hour by helping to avoid drinking on an empty stomach. Hors d'oeuvres are usually passed but can be set on a sideboard or table and can be as simple as toast points with pâté de foie gras, egg salad, or bruschetta all the way to giant bowls of chilled lobster claws or stone crabs and a mustard rémoulade sauce. A raw bar can make for a popular hors d'oeuvre station. Sometimes it is as simple as a crudité of farm-fresh vegetables and a dipping sauce. Your knife skills will really make a difference, so go for finer, more elegant cuts, no stalky cuts that remind you of a fourth grader's lunch. Beggar's purses are a fun way to introduce a mouthful of flavor in an attractive, easy-to-eat

FIGURE 11.1   HORS D'OEUVRES IN ST. BARTHS AT THE MILLENNIUM.

FIGURE 11.2   SERVING COCKTAIL HOUR CRUDITÉS.

package. Garnishing the platters is a little less restrictive here, and you can definitely cost-average in some whimsy over function.

Hors d'oeuvres are your chance to make a first impression and win over the diners with a variety of flavors and a beautiful display. It really does pay to put some effort in. It is fine to use mature flavors from your

FIGURE 11.3 LOBSTER FLOWN IN FROM LEGAL SEAFOODS IN BOSTON.

refrigerator. Leftover prep can be made into great flavor combinations in a beggar's purse; however, do not confuse a tiny encore with laziness or cost saving. We do not sell food in the private chef world; do not use anything that is not still fresh or totally delicious. There can be a lot of leeway in an hors d'oeuvres spread; the cost can go from $40 to $400 or $4,000 really fast, so make sure you and the boss are on the same page depending upon who they are entertaining.

Amuse-bouches have become popular in the last few years—or should I say they have become popular again. *Amuse-bouche* translates as "to wake up the palate," or at least stimulate it. They differ from hors d'oeuvres in that they are usually eaten at the table, are one to three bite-sized pieces served as a peace offering directly from the chef, and are entirely of the chef's choice. Amuse-bouches usually correspond to the night's theme or menu and are usually not one item; for example, a raw oyster or a single, chilled cocktail shrimp has a variety of textures and is flavorful. They are usually served on a bread plate to the side of the

FIGURE 11.4   RAW BAR.

charger (or show plate) immediately after the guests are seated. They are not as substantial as an appetizer, and there is no need to add a garnish.

Whether you are serving an hors d'oeuvres platter or single amuse-bouches plate, temperature, color, and variety count; however, as the diner's first impression of the meal to come, flavor and texture still rule.

CHAPTER 12

Dinner

For most chefs in the private service world, dinner is the main focus of the day. We start dessert projects, trim proteins and begin to plan plating designs early in the morning and work on dinner related prep throughout the day. Breakfast and lunch seem more utilitarian in comparison. Dinner will last longer and is when the fine china starts to show itself.

Many of my bosses will shower and dress for dinner in their own homes. I have seen more than one Mrs. arrive to the table dripping in jewels on a Wednesday night just to have dinner alone with her husband.

Dinner will more often than not be a three-course affair, beginning with a starter of a soup, salad, or an appetizer and followed by an entrée consisting of a protein, starch, and vegetable with sauces served on the side to give the boss more control over dinner. Finally, a dessert, fruit platter, or cheese board will usually top off the meal. Dinner can expand to multiple, smaller courses on special occasions. It can be served French style (plated in courses), Russian style (where the butler serves portions off of a silver platter at the table), butler or British style (where the butler holds a silver platter and the guests portion out their own at the table), family style (where platters are put right on the table),

FIGURE 12.1   YES, A TABLE THIS LARGE FITS ON A YACHT!

FIGURE 12.2   A BRUNCH FIT FOR MARTHA STEWART.

or Buffet style (where the foods are lined up on a side board for self- or assisted service).

Truth is, throughout the year most households will use a combination of these depending on the occasion. The chef will be responsible for tidying up the platters on the buffet between rounds. The last pass through should rival the first.

Careful attention should be paid not only to the choices on the menu but also to the portions. Whether or not they talk about it out loud, most rich and wealthy people hate waste and seem to prefer smaller portions with the option of seconds. This keeps those with poor self-control from overindulging or putting pressure on a light eater to finish a portion larger than they are comfortable with. Foods should be plated cleanly; sauces are served on the side not only to put portion control in the hands of the diner but also to prevent sloppy plates or a sauce's intrusion into a different texture or color on the plate.

Hot foods are plated on plates warmed in the oven for 20–30 seconds beforehand. The proteins, starch, and vegetable are generally plated close together or touching to consolidate heat. Entrée plates should not be huge, and garnishes, if needed, should be edible or functional in some sense. Nothing on the plate should come back—not a bone (unless a proper chop), not a sprig of parsley, not a piece of gristle or connective tissue. We trim and butcher the tough and chewy stuff away. We offer only fork-tender portions that are easily broken down. More likely than not the protein is sliced on the bias into pieces that can be eaten in one or two bites. This ensures the chef has inspected the doneness of the proteins and also takes most of the work out of cutting for the diner.

Salads, cold appetizers, and desserts are to be served on cold plates. If they are put in the freezer in advance of the meal, by the time they are plated and served they are cool to the touch. This helps to stop vinaigrettes and dessert sauces from running and keeps the lettuce from wilting.

Socially, there is a big difference between a lunch date or lunchtime meeting and one held over dinner. Dinner is more intimate than lunch and is at a slower pace with more time and money invested. This is a time for good conversation, and the outcome of a dinner engagement is expected to be more lasting. Dinner is a meal that requires balance with few peaks and valleys. The meal should also be easy to digest.

When you are planning a buffet, it is important to forget what you know and remember that you are not selling food. Abundance is king here. A buffet for a hotel will have dozens of crappy, starch-based, inexpensive salads that no one ever dreams about. You put your cheaper filler items out front and make the customer overfill their plate before

FIGURE 12.3   COLORADO RACK OF LAMB.

FIGURE 12.4   VEAL MORELS.

FIGURE 12.5 VALRHONA CHOCOLATE SOUFFLÉ.

FIGURE 12.6 BEGUINE CHOCOLATE TERRINE.

they ever hit the good stuff. The carving station guard then doles out a slice or two of the protein and makes you beg for an extra slice, like you aren't paying full price for it.

In private service, we do not mark up the cost of food. We are paid to hunt and gather for the boss and then prepare the food. Nothing is more awesome than walking up to a sideboard with piles of Maine lobster out of shell swimming in drawn French butter, chateaubriand finely sliced and drizzled in white truffle oil, and a sea of sides that you actually can't wait to eat. The trick is to not have the platters be too big or deep and replace them with fresh backups frequently. Encourage the guests to have all they want. A buffet should be a celebration of choices and indulgences. I do know of one restaurant that does this. The Nordic Lodge in Charlestown, Rhode Island, serves unlimited lobster piled up, great steaks, all the good sides you could want, and a Häagen-Dazs sundae bar. It just charges a lot and gives you a 2-hour limit, but there is a line around the block year in and year out. Nordic Lodge is so successful because anyone would want to experience that every now and then. If your boss has unlimited funds, then "make it rain" on these occasions. Just make sure that you do not have piles of food left over. There is an art to this, and it begins with good communication with your service partners.

Want to know how to serve a great dinner? Learn how to be a great diner. A cocktail upon arrival will start to erase the stress of travel and the challenges of the day and will aid in relaxing your guard toward your fellow diners. A hors d'oeuvre or an amuse-bouche will begin to wake the palate. When you arrive at the table, you settle into your new surroundings; placing the napkin on your lap signals your readiness to begin the adventure. Think of it like buckling your seatbelt when you get into the car. Next the wine is poured; partake, it will relax you, let down your inhibitions (important for a stimulating conversation), and aid in the digestion of the meal to come. Red or white? Like the French say, "Why can't we have both?" Take time to look your fellow diners in the eyes, as this initiates a bond. Take some time to look at what they are wearing, since many have taken the time to show you respect by grooming and dressing to present themselves to you at their best. Smell the flowers and anticipate the chef's hard work and creativity. Submit to the whole event. To get the most out of the art of dining, you must completely submit to the experience and live totally in the moment.

Freshly baked hot breads arrive in baskets with side dishes of piped French butter. The butler discretely slides a hot bowl of soup on an underliner from the left on the top of your charger. People eat at a relaxed pace and share a few topical discussions, usually on topics in which they are more educated. Many people share in the conversation and allow for everyone to finish the first offering without anyone feeling as though they are stragglers. The butler silently removes each diner's soup bowl, underliner, and spoon from the right without any interruption in conversation; he is like a trusted ghost. The conversation warms up with new topics. The butler then arrives with a chilled appetizer while the conversation slides from topics of current events to ideas and solutions. Eleanor Roosevelt once said, "Great minds talk of ideas, average minds talk of events, and small minds talk of people." This is a good guide to dinner conversation. Religion and politics are too personal for mixed audiences and divide more than unite; they are best left for more monastic groups. At the end of the appetizer course, the diners rest their used utensils on the plate at a 45° angle discretely to signal that they are finished. There is not usually a spoken communication with the butler unless instructions are out of the ordinary. The butler will read the boss's plate as the master guide to the pace of the meal and will report to the chef when each course is 60% finished. The butler removes not only the appetizer plate at this point but also the charger with it. If any silverware has been misused or dropped, he will resilver from a small tray. Wines and waters are refreshed throughout the meal; drink as much water as wine when drinking reds.

Entrées come out two at a time; a server never carries more than two plates to the table. On a yacht or large estate, many staff will carry two plates each and bring them to the edge of the dining room just out of sight; the butler will then take them out of their hands and place them at the table. Continuity of service is important as not to break the concentration of the diners engaged in their own experience: the smells and sights of a candlelit dining room; gorgeous, fragrant, fresh-cut flowers; the artisanship of the fine stemware, china, and silver. Fine music and great conversation almost pale in comparison with the unmistakable glow of a candle on a white tablecloth behind a paper-thin leaded crystal Bordeaux glass filled a third with a claret (fine Bordeaux red wine) that hasn't seen fresh air since you were in grade school.

The hot, well-seared meat, the lightness of the duchess potatoes, the crisp brightly colored vegetables, and the sauce—that sauce! Men have gone to war for less! Each bite melts in your mouth, and you chew slowly and keep each bite in your mouth until you get butterflies in your stomach and can't take it anymore. Nature and artist have come together in their finest union. No one talks for minutes on end. You take that claret and raise the edge of the glass to the bottom of your nose and just breathe it in, body and soul—drifting a Lamborghini around an S-curve with all four wheels spinning has got nothing on this.

Dinner is cleared; you are viewing your fellow diners with a sense of afterglow in your eyes. All savory accouterments and superfluous hardware are removed from the table (bread, butter, salt and pepper, bread plates), and then the table is crumbed. A cordial cart or tray is brought to the table. A cognac, Sambuca, or fine port seals your passage into a state of bliss. The table is properly silvered for the final course. This is the time where you may reminisce over past, shared experiences or have more intimate conversations. A gorgeous, plated dessert or potent cheese board is brought to the table. Someone will inevitably comment on the offering using the word **heaven**. When the course is finished, the group will usually retire to another room, cordials in hand. After a fine, well-paced meal, you should feel relaxed and alive. You should be able to make love, not unbutton your trousers and have to lie down for a while.

That is the dynamic of the dining room: you cannot rush it, and you cannot drag it out. Butler, host, and chef must all work as one. There should be a gentle tension carried throughout the meal like water through a pipe at a steady, medium flow.

CHAPTER 13

Show Time

Have you ever read one of those celebrity gossip magazines and seen a picture of a $30–50 million wedding somewhere in the world and said to yourself, "Who does that?" or "How do you even pull off something so extravagant?" In short, the answer is: someone who can afford that and has a crack staff of private department heads teamed with an army of hired temporary staff and professional coordinators. It is quite a sight to see a billionaire actually spending money, and by that I mean consumables or one-time use items and experiences. Normally, when they buy a painting or a home, it will increase in value. Keeping in mind that time is money, a private jet can actually make them money over time. Cost be damned when the boss wants to have that once-in-a-lifetime experience and it really is something to be a part of.

Cooking for a billionaire demands some serious entertaining from time to time. This can be terrifying for those who are not used to high-volume production or managing a team. This can also be terrifying for those not used to the stress of cooking for high-profile guests in an intimate setting. Sometimes show time is not about the size of the event but the importance of it. Sometimes the boss is trying to have a private in-home meeting to decide whether 50,000 people go to work next year or not. If you blow the dinner and it becomes a disaster, the guest may think, "If the boss doesn't have their own home under control, how can they handle the subject at hand?" No pressure there. Once you get your confidence in check, it is all about planning and then pulling out all the stops!

The first step is to sit down with the boss or leader of the event and figure out the goals and purpose for the occasion. This is important because sometimes it may not be obvious. One instance that comes to mind is when I was doing a trip on a yacht in St. Barths for one of the very richest families in the world whose many holdings included a very well-known department store chain. Expected on short notice, for a cocktail hour, was a very high-profile American TV presenter (also a billionaire). Most of the crew were South African and therefore not familiar with American television or the presenter, but the Americans and Brits were like, "Wow! This is a big deal." The South African chief stewardess approached me in the galley and informed me that Mrs. said, "Don't go overboard; it's just one of our employees." Wow again: I am a nice boy from New Hampshire, and this is on another scale. It was hard to hold back, but I knew Mrs. could tell if I was showing off, and she wanted to send a message of dominance. Solid but standard fare was served.

There are a lot of subtle messages sent in the social strata of the wealthy. It is not your decision to give 150% effort when your boss is not trying to impress someone they don't really respect that much. Yes, every-thing should be excellent and at its best but also within boundaries. Having the boss's son's little league team over for a barbecue may not require a Kobe beef steamship round; really good hamburgers are prob-ably just fine. In-home civic events can be nice, but I have seen the boss host without even coming downstairs to making an appearance. Donating a fleet of cardiac ambulances and allowing the event to be thrown in his home was considered enough.

I do believe that the worst thing you can do is to underwhelm your boss or their guests, but keep in mind a common theme in this world is that the rich people hate both waste and being taken advantage of. Hijacking the event to stroke your own ego on your boss's dime is also considered a high crime. It is best to keep yourself out of the equation and focus on the execution of the task at hand. Once you understand where your boss's head is at and you know what they are trying to accomplish, you can move forward. As usual, if your boss wants to really make an impression on someone, it all comes down to being a good detective and finding out if the guests have any individual requirements and anticipating them with confidence, not reacting to them during the actual event.

Microsoft cofounder Paul Allen, worth $46 billion, is a big lover of hamburgers; and if he is coming to dinner find out if he prefers Kobe, Lobel's, or Whole Food's grass-fed beef. It is always good to know if someone has allergies or is gluten intolerant ahead of time. Aristocracy's valets, who know their employers preferences, and your boss's personal assistant, who usually takes care of the invitations, are often your best resources for this kind of information, which makes things a lot easier.

Now you are ready for the next stage: planning the full menu and engineering the event. For the big events, write down the who, what, where, and when, and check the weather forecast if the event is close enough. Create a menu and get it approved. Start working on the portion requirements and make order lists. Walk through the area and get a feel for where things are going to be set up. Satellite bar and hors d'oeuvres stations around the event are good to keep the guests from bunching up in one area. Wealthy people like to break off into smaller groups around an event and then drift during the evening. If you have only one bar area, people tend to feel held hostage there out of practicality. Have a central full-service bar, and add a few small stations in other rooms that may serve only champagne, wine, and maybe a good scotch. Maybe another will serve just craft beers. People will congregate where they feel most comfortable. It's nice to be hiding out in the scotch tent with people of your kind (single-malt people), or maybe you prefer the champagne clutch. The crowd lovers will orbit at the big bar. Creating flow and attractions throughout the event is the key to people mixing. Ideally, everyone should meet everyone else throughout the event, even if just for a moment. Parties are not designed just to feed a bunch of people efficiently; they are to promote social interaction within a group.

Good music that supports the theme of the event is another key element. Live music is always best; it is just something that makes the event unique. Regardless of the choice of music played, they are the only people in the world with that mix of musicians that night. If the event is big enough, staggering multiple acts throughout the night in different locations can create a sense of adventure and encourage the guests to change rooms throughout the event. A dance band at the pool, chamber music in the library, and a flamenco act with dancers in the parlor will create interest. Stagger but overlap the performances. Have a live act or two do one-time shows, such as a belly dancer or an aerial acrobat, to create a sense of urgency to flow to the center of the action. I once worked for a family in Palm Beach that brought in black swans to swim in the pool, where they had a temporary bridge built over the pool. They had flown in a troupe of topless beauties wearing full headdress, straight from the Moulin Rouge in Paris, to roam the party and then do a full show on that bridge.

Allen has an annual New Year's Eve party in St. Barths aboard his 418-foot, $200+ million James Bond villain–level gigayacht, the *Octopus*. His headlining entertainers for the party have included such acts as Bon Jovi. At the same time, a few islands away, I was doing a holiday trip on Aristotle Onassis's legendary 325-foot yacht, the *Christina O*. One night, we held a barbecue for our billionaire charter guest and 31 of his closest friends and family. The staff all dressed like islanders, we cooked local cuisine on deck, and hired a 15-piece steel drum band to create a different, yet still unforgettable, evening.

Chances are you, as chef, are not going to be burdened with contracting and coordinating a New Jersey–based supergroup to be choppered onto a seven-story, celebrity-filled yacht. However, you may be a part of or in charge of catering it. Both Allen and Bon Jovi will have to be fed, as will the staff of close to 100 as well as oceans of the rich and famous. You had best enjoy the logistics and the challenges of doing this 1,500 miles off the mainland all while dealing with international customs.

The trick at this level is breaking the symphony down into sections, breaking the sections down into individual musicians and then giving them the right sheet music. Staff meals, meals for vendors (e.g., entertainers, photographers, outside security), hors d'oeuvres, proteins, side dishes, and pastries have to be planned out with designated time slots and final consumer destinations. Then it is time to establish how many staff you will need and if you will need to supplement your team with freelance labor. Once this is all in place, you will need to make assignments as to who is responsible for what. This all sounds like common sense, but during a busy holiday week the boss may have several sizable events hedging the superparty. Where to store the incoming foodstuffs and the fully prepped foods as well as the par-cooked and fully ready-to-go foods can be overwhelming on its own.

One thing I learned at the Culinary Institute of America was to always be bigger than your problem. If you have a yacht with a finite amount of cold storage and the boss wants to host their daughter's wedding onboard, don't waste time staring at the cooler wondering how everything is going to fit. Rent as many 150-quart coolers as you can safely fit in the areas of the boat that are not being totally utilized that day (I once put one in my shower). The regular rules of life should be used only as guidelines when the boss wants to go all out. As we chefs always say, "Get it done, get it done right, and get it done right *now*, even if a few villagers have to die." If the island does not have a rental store, you can also buy the coolers and then sell them after the event in the marina at half price the next day. Mariners love bargains: used only once.

There are times where the algorithm changes and you are now out of the operating limits of your team. These are the times that you need to call in an outside team. This is where the caterers come in. You will encounter two catering scenarios. In one, you are having an event held on the boss's turf and you are there to guide the caterers to make their job easier and make sure that they have what they need to succeed is the first. Your presence will put some accountability into the mix so the boss's equipment doesn't get abused or go home with the caterer. This situation happens when the event is just simply far outside of the scale of your equipment and team, such as a 150+ person wedding. The scenario can also be that the yacht is arriving from overseas into Palm Beach and the boss wants to have 75 people onboard the day after you arrive. There is simply not enough time to source out the foods and prep. The yacht staff will be exhausted from running a 24-hour schedule for days on end, so the boss can easily make a few calls and you will oversee the event without participating in it.

FIGURE 13.1   ON A YACHT, GOOD SPACE IS AT A PREMIUM;
SOMETIMES YOU NEED TO LAY YOUR CLAIM.

The other situation will be when you augment your staff or menu with resources from a caterer. There is good and bad with this scenario. It is always fun to bring in a few outside people for an evening. It is a novelty for us to share our crazy, seldom seen world with some bright-eyed, bushy-tailed people who are not jaded to the grandeur of our private and privileged world. It reminds us of when we first were exposed to it all and how lucky we are at times. The bad part is that the average food-service worker is not used to the way you have to operate on a yacht or in a jewel box of a home. Fingerprints show on everything, and you cannot wear shoes that have fine-ground dirt and pebbles imbedded in them (if they have been worn outside, they have them) on the $198,000 carpet or white marble floors. Also, an ashtray can cost $4,000; they are easily pilfered or, even worse, broken! It is important to respectfully have a little meeting that does not make them feel bad but that will orientate them to be deputy members of the staff for the night.

The bad part about having a caterer supplement your food with extra dishes is that most private chefs use the finest ingredients in the world: $250/gallon olive oils, butters that cost more than fine chocolate, and garlic that costs $5 a head. Your local caterers, as fine as you may want to think that they may be, are not using any of these ingredients. Their business is to buy low and sell high. You can request prime meat and get it—even cooked properly—but the devil in Michelin quality cuisine is in the details and in the chain of ingredients that makes up a dish. Everything that goes in that dish must be the best. Caterers focus on what they can charge you for and agreed upon in the dish. For example, in prime New York sirloin with truffled duchess potatoes, the steak and truffles will be real but the butter going in the potatoes will be a no-name, watery, $2/pound brand instead of a French cultured butter, the kind that will make you now understand all those love songs on the radio. Every time I have ever supplemented my fare with a caterer's, the boss and more culinary savvy guests can always tell my food from the caterers and my stuff goes first. It is not because of my culinary superpowers; it is because of the ingredients I use. I think a catered affair is one experience, and there is a time to use one and there is a time to hire extra workers in on your turf and burn the candle. Sometimes you just have to mix different qualities of cuisines and bite the bullet.

No matter how you do it, though, communication is the key to success. A 3-minute pep talk with front and back of the house can really get everyone on the same page and bond a team. Be a leader and answer all questions on the fly, from anyone, quickly and politely. And let everyone know what is expected of them in advance.

There is the other show time, where it is not all about logistics. This is the one I mentioned earlier when the boss is having an icon visit or is under stress to have something go right in home with visitors that count. I, as an American TV gross consumer, get excited about celebrity visitors and clients. I used to try to get the rest of the staff pumped up for this. Most of my coworkers are not from major television-producing or celebrity-producing countries. Over time I got the point that my coworkers did not care who the guest was, just that the service was good. It is the better attitude to have. Celebrities face enough paparazzi on the streets; they do not need people freaking them out in private. Just go about your job and make them feel at home, and your boss will appreciate this. When it is important, go over your checklists several times, make sure you have alternatives, and then go about your business. Do not take on your boss's stress and create a pressure cooker environment among the staff. The guests can feel what the staff is feeling, and if the boss is thinking about what you are feeling, they will not have their head in the game. On important days, meet with your service staff and make sure that everyone has what they need. Go over checklists together, be supportive, and then just go about your business. You are all trained, experienced professionals.

CHAPTER 14

The Houses

It is really something the first time you arrive at the gates of a spectacular residence and they open for you. Rolling down the driveway to a blank canvas for your art, you face a new adventure and a microcosm of characters you are not likely to forget. Working in one of the boss's homes is different from working in their other assets. It is more personal, and operating within the power structure can be much more complex; however, the actual cooking can be much easier than on a yacht or private long-range jet or even a private island.

There are two types of private chef jobs in estate work: live-in and live-out. We will not be dealing with hourly or part-time workers or basic cook positions here but full-time or freelance professional chefs only. When you are freelance and flying in for a set amount of time, housing should be provided and will be in the form of a private room in the main house, on the property, in a nearby home, or in a hotel or motel in the area. A trailer in the woods 20 miles away is unacceptable, and you cannot expect your own house or luxury apartment either. However, sometimes luck is on your side, and in this business oftentimes you win. In bigger programs, you will probably get your own apartment or even a nice house nearby. You can also expect the use of a vehicle, and staff cars can run the gamut as this is a totally unregulated business.

FIGURE 14.1 TYPICAL COUNTRY ESTATE HOUSE.

The standards are an SUV or minivan, but a brand-new Range Rover, Jaguar, or S-class Mercedes Benz may end up in your hands for your tenure. Enjoy and treat it well, because it is a great perk. Staff cars can be available to department heads for personal use. The unspoken rule, though, is usually not to take them on long overnight journeys without prior consent, and if so you would pay for your own gas. On those occasions, it really depends on the boss and how much they like you.

Most estates are in luxury areas where rents can be prohibitive on a staff member's budget, but the boss will want some people to live on the property full time to be the eyes, ears, and 911 dialers. Live-in, loyal staff are figured into the security plan as it is preferable to have some of the staff around than a private army roaming the grounds 24/7. Not having to pay rent is nice, although you may have to pick up a landline, Internet, or a cable bill—but realize that nothing is free. When you live in one of the boss's assets, other staff will come by unannounced for many reasons over time, such as to take photos for insurance, for maintenance, or to pick up something stored in the attic or basement for an upcoming event. Anything perceived as out of line will make it back to someone in charge at some point. A few bottles of wine or a single-malt

FIGURE 14.2   SOUTH FLORIDA OCEANFRONT MANSION.

whiskey sitting on the counter waiting to be gift-wrapped and brought to a party the next day can turn into, "I think the chef is an alcoholic; there is booze everywhere."

Of course, it is great not to have a commute, and if the staff are cool you can have some memorable times; however, the politics in an estate job can be brutal. If you are loyal staff member and give up a sizable chunk of your life in service to one of these elite families, there is a pretty good chance that you may end up with a substantial severance package upon retirement or even be entered into the will. I once knew a girl who started as a cook on a boat; the owners grew fond of her, and the Mrs. began to look at her like a daughter. They eventually asked her to become the estate chef for them as well. Just 10 years later, this untrained cook was pulling down a $300,000 salary, received a new luxury car every other year, and was able to gift the previous one to whomever she pleased. Eventually she received 1% of each land deal that the family made. She is now in her fifties and is a millionaire several times over. Doris Duke, the richest woman in the world at the time of her death, had the bulk of her fortune land into the hands of her barefoot, gin-blossomed, uneducated butler, who after her death started dressing up as her. I tell you these things not to create motivation but to let you know that there are some very ambitious staff members out there and to be very careful to stay out of their way. If a butler has the ear of the owner at all times, they can easily put a knife in your back. You will never know what they said to cause your demise, but I imagine that it would be something based on a true small infraction that was exaggerated and now you seem like a risk to the owner. This is definitely a CYA (cover your ass) industry. It is best to stay under the radar and just do your job and enjoy your craft. Do not get involved with household politics.

Chefs have jobs of both discipline and creativity. It is rare for a good chef to stay with one family for decades, and a good chef will want new creative challenges. A good butler or head housekeeper or grounds-keeper may want to run one well-oiled estate for a whole career.

When you live outside of the property on your own dime, you are really buying your freedom and your privacy at a very reasonable price. You can have whomever you want over and live life on your own terms without repercussions. It is very important to leave extra time in your commute as accidents and unexpected construction can ruin the boss's morning; remember that there are no excuses in private service.

There are other plusses to working in the boss's home. You usually end up with a bigger kitchen, which is generally not a hot kitchen and is often very well appointed; you will have better storage options and consistent purveyors who will know your expectations. You can also have a life in these situations. If you have a family you can have them live with you in some situations, but this is rarely an option on a yacht or private island. You can also make friends in the community; you can establish a relationship with a local house of worship, join a local softball league, or just have a local pub that knows your name and your drink.

One of the most important things to remember when working in the boss's home is that it is the boss's home. You are not in a restaurant, the boss is not staying in a hotel, the boss is not on vacation: the boss is at home—*their* home sweet home. It is sometimes difficult for most people to get their heads around the fact that a 1,000-acre hillside retreat with a palace bustling with staff can be a home, but it is. The boss

FIGURE 14.3   EXCURSION THROUGH THE SAHARA.

needs to feel relaxed and free in their home just like you want to in yours. The boss should feel free to leave valuables anywhere they wish and needs to be able to trust their staff with not only their assets but also with their secrets. They need to feel confident that their staff is not out all over town telling stories of what really goes on in the castle on the hill. Be cool. The townsfolk live for stories and gossip and will pull up a chair to hear your tales. Nine times out of ten, they will even buy your drinks while you gain attention and notoriety, all at the expense of the man or woman who pays your bills. What happens on the estate stays there, and privacy and trust are paramount. Your boss is just as susceptible to embarrassment and feeling betrayed as anyone else.

You must also relax to some extent while dealing with the boss, as people often mirror the attitude that you give. Smile, ask how their day is, and give them your full attention. Never argue; a simple, "Will do," will suffice. How would you want people to treat you in your home? Well, probably like it is your home. Listen to what the boss says. If you say, "Hello, Mrs. Jones," and she says, "Call me Muffy," then call her Muffy—except in front of company or outside of the home. If someone wants you to use his or her name at home, do it. On the other hand, if you are collecting a paycheck from someone, you are not their friend, so be careful not to mistake a genuine liking of you and a comfort with you with being on the same level. Respect is everything in this business. When you complete your employment with them and they invite you on vacation with them, you can now consider that you are actually friends at that point. Many of us have been on charter with someone and they invite us to come visit them at a later time; it is another perk of the business.

A full-time chef is expected to provide agreed on meals and provision, to maintain kitchen equipment, and sometimes to serve as well. Other duties may include cooking for, grooming, and walking pets; driving children to school; doing airport runs; delivering packages; pet sitting; planning events; maintaining an herb garden, household, or the boss's personal shopping; supervising workers and delivery people; and being the purser (cash and receipt person). Days off are to be determined by the owner. Any personal time off should be requested far in advance and is not usually guaranteed unless a direct family member is in a dire situation. You may have to work sick if it is not contagious.

The chef is a department head in a home and answers to the owner, estate manager, butler, or major domo. In some households, the chef is in charge so the chain of command should be established early to avoid ruffling anyone's feathers.

CHAPTER 15

The Yachts (Rock Star Chefs)

If James Bond would have chosen another career path, there is no doubt in my mind that he would have been a superyacht chef. Other than a secret agent with a license to kill, I cannot think of another job where you slide into all of the world's most exciting ports of call with a American Express Black card, $10,000 in your pocket, and a mission to carry out for one of earth's wealthiest residents—with the very real possibility that they are a world-class villain.

Nothing has affected the private chef world more than the quintupling of the private yacht business in the 10-year span between 1998 and 2008. Real chefs finally got the chance to work as artists with billionaire benefactors on a global scale: flying or cruising from country to country to meet up with a billionaire who is on an adventure, not just eating at home.

The proliferation of the private jet industry and an upswing in the development of long-range private aircraft now made it possible for the boss to meet up with us anywhere on a whim.

The advent of the advances in brand-new industries based on communications such as cell phones and the Internet blew up the stock market and created a global market accessible to anyone with no one

FIGURE 15.1   YACHTS AT A HOUSE IN THE CARIBBEAN.

FIGURE 15.2   YOUR UNDERWEAR MONEY HARD AT WORK: LESLIE WEXNER'S FOOTBALL-FIELD LENGTH GIGA-YACHT "LIMITLESS." MR. WEXNER'S BILLIONS COME FROM HIS CLOTHING AND RETAIL EMPIRE THAT INCLUDES VICTORIA'S SECRET BRAND.

to yet oversee it. All of this happened in the shadow of the new millennium, which brought both fear and optimism to many brand-new, overnight millionaires. The stock market became the world's largest online casino, which led to the trickle-up economics that made many of the world's superrich now ultrarich. It was no longer that impressive in the jet-set world to have a few hundred million dollars; there was a new word now going to cross the lips of nearly all Americans on a regular basis: *billionaire*!

FIGURE 15.3   SUPERYACHT COMPLETE WITH CHOPPER.

FIGURE 15.4   IT'S NICE WHEN THE RULES DON'T APPLY TO YOU.

In 1998 there were roughly 1,700 luxury private motor yachts over 100 feet in operation around the world. Most were exactly that: around 100 feet and each with a crew of four.

Whoever was dating the captain would come along for trips as the cook for a crew of five. A few billionaires back in the day would have the

FIGURE 15.5 289-FOOT *MALTESE FALCON* DWARFING A MEGAYACHT.

random 130–170 footer, and a sultan or two would have a monstrous commercial ship or military vessel converted with a luxury interior; however, they seldom roamed far from their home ports or Monaco. There was no professional crew at that time other than a licensed captain who usually did their own engineering as the boats were used largely as seasonal harbor hoppers: north in the summer, south in the winter.

Within a few years—and no one knows why—the ultrarich went nuts and all at once started ordering huge custom yachts faster than the yards and designers could make them.

The captain's wife or girlfriend could no longer handle the new demands of global travel and feeding an international crew of 17. If you were a trained chef, drug free and not a complete alcoholic at this point, you were about to be sucked into a vacuum of opportunity beyond your wildest imagination. If you could cope, you could write your own ticket. We would become the highest paid, most perked chefs on the planet. It was beyond the Wild West. It was not out of the question for a chef to have their own exotic car trailered into the Hamptons for the month because they missed it and did not want to drive their rental car on their time off. With tips, some freelance chefs were raking in $1,000 plus a day with no expenses. The vast amount of the work was paid

FIGURE 15.6 JIMMY BUFFETT'S *CONTINENTAL DRIFTER* IN THE CARIBBEAN.

in cash and untraceable. Boss and crew member egos got blown out of proportion. Owners pushed for the impossible, and captains feared saying no.

Just when it got to be almost too much to bear, though, the stock market crashed in October of 2008. It was a mixed blessing. It was the end of an era of excess, but it became a tough market for chefs for several years. Now that things have leveled out to a reasonable pace, a new yacht chef can now have a career that begins with experienced professionals around them and slightly more realistic expectations from their owners.

You must get your head around some terms and numbers before you go betting the farm on a 007 chef career. It is not for everyone. The term *yacht* is defined as a pleasure craft 10 meters (33 feet) and above. Entry level for a full-time chef position is on a megayacht, defined as a luxury private motor yacht 30–49 meters (99–161 feet). A superyacht is generally considered a private yacht of 50–74 meters (164–244 feet), and a gigayacht is any private yacht 75 meters and up (247 feet). This is only a partial guide, as length is not the sole mitigating factor. Tonnage, design, styling, and interior can make a difference from two 350-foot yachts having a difference of $100 million in the asking price. That is *not* a typo.

FIGURE 15.7   GERMAN BILLIONAIRE REINHOLD WUERTH'S 280-FOOT
$100 MILLION GIGAYACHT *VIBRANT CURIOSITY* FEATURES A CREW OF 26.

Yachting is the most expensive thing a human can spend their money on. Mansions can cost a pretty penny but are much cheaper to run and can be mothballed if needed. They will go up in value over time if they are well maintained. A private jet has an average lifespan of over 30 years and can live in a hangar indefinitely with no one living onboard. It will actually get you to emergency meetings globally in one day to stop a crisis and potentially save tens of thousands of jobs and billions in investment. There is nothing you can do on a yacht that speeds up business or makes a business more efficient. Yachts are profoundly wasteful by nature. Most do not often exceed 15 miles an hour and consume vast amounts of fossil fuel just moving them around and making electricity. They require full-time, live-aboard crew and need to be cleaned and polished 365 days a year. No two yachts are alike: standard parts for most of the boats do not exist, and parts must be fabricated.

Why would anyone own one? Because the one thing they can do is inspire man—in craftsmanship, form, and function. They create the most indelible memories and offer freedom unmatched by any other possession on earth or beyond. Some of the yacht owners who have private space programs still need government permission to operate them. Yachting is about the most freeing activity you can get involved with. If you don't like your neighbors, move. If the dolphins jump higher in the next bay, go there. If you are on the way to a Caribbean destination and your friends invite you to have dinner on their yacht at a nearby island, call the bridge and tell your captain, "Turn left; change of plans!" Try that on a cruise ship. What separates yachts from estates in cost is the upkeep. The general rule of thumb is that a yacht will cost about 10% of its value in annual operating costs, so if you go out and spend $10 million on a boat, in 10 years you have spent $20 million for the experience. Most yachts are used an average of only 12 weeks a year and are considered a fifth home that can be anywhere in the world on short notice. As of 2014, there are about 5,000 luxury private motor yachts on Earth to service seven billion humans. Those numbers should be staggering to you. Only a small percentage of humans will see one of these yachts in person from the outside in public, and even fewer people will ever see the inside of one. It's not just the fineness of the joinery and stonework, but it is all handmade and built into the vessel by the world's top designers and craftsman. Ralph Lauren designed the interior of a yacht once, and it was considered a compromise. The designers at this level are names that the middle class will never know. Tens to hundreds of millions of dollars are spent on interiors. Entry level into this world is about $8 million for a 100-foot family cruiser, and I personally watched the building of a 600-foot gigayacht in Germany rumored to be the world's first billion-dollar yacht. An employee on the project told me over dinner that he traveled the world to design the audiovisual system for the whole vessel and dropped a cool $50 million on sound alone.

When the boss is not using their boat, they sometimes let it be chartered to other titans of industry to help offset some of the operating costs and give the crew something to do and earn some extra money in the form of tips. Charters start at around $65,000 a week for a smaller family boat, and that does not include fuel, dockage, food, and drink or tips. So figure $100,000 a week with the average charter being two weeks. Many of us on the bigger end have done charters that top $1 million a week. Of the 5,000 yachts out there, only about 100 are gigayachts and the sky is truly the limit on spending and how crazy things behind the scenes can be.

FIGURE 15.8   RUSSIAN BILLIONAIRE ANDREY MEINICHENKO'S 390-FOOT $300 MILLION YACHT *A*.

FIGURE 15.9   IF YOU HAVE A $17 BILLION FORTUNE AND FEEL LIKE DROPPING NORTH OF $80 MILLION FOR A FAMILY GETAWAY MACHINE, YOU CAN HAVE THE *M/Y AZTECA* LIKE MEXICAN BILLIONAIRE RICARDO SALINAS PLIEGO.

FIGURE 15.10   Chef Neal ready for a night off in Boston.

FIGURE 15.11   View from a gigayacht in St. Martin.

What is it like being a chef on a yacht? There is a world of difference between being a chef on land and one at sea, even if both are private situations. You are not just a chef in these situations but also a mariner and part of a crew with multifaceted roles. The cooking and menu planning are not radically different; however, you will now have a second captive audience in the form of a hard-working crew that will expect a very well-prepared, healthy nutritious lunch and dinner at noon and 6:00 p.m. You also may not have any help in the form of scullery help and may actually have to lend a hand to an overworked stewardess staff by rinsing plates and loading the dishwasher. Chefs on yachts are expected to source and acquire provisions and handle payments, tips, and bookkeeping. They must arrange, store, and monitor the condition of the provisions, which may have to last for weeks on long voyages and when traveling in remote locations. Yachts make very good use of every inch for storage: we use the underside of the cushioned part of settees, chairs, and mattresses on the bunks as well as the bilges as our secret hideaways for everything from linens and paper goods to canned goods

FIGURE 15.12 ONE OF THE SALONS ON A SUPERYACHT.

FIGURE 15.13 STATEROOM ON A GIGAYACHT.

and bottles of wines. Space on yachts is always tight as we often have acres of square footage but much lower headroom than in the boss's house. Most of the boat will be used to justify the boss's seemingly impossible investment in the journey: room after room of the finest handmade, built-in furnishings with 27 layers of hand-rubbed finish next to marble that is now obsolete because your boss bought it all to demonstrate that they now have something no one else does.

The crew live together in a section with a common area called a *crew mess*, where everyone takes meals and can relax and watch movies. Most yachts will have decent AV equipment, but most crew today will retreat to their bunks to escape into their own computer to watch something of their own choosing. It is fun, though, to sit with your crew and enjoy a movie or a midday meal while catching up on what is going on around the vessel or in town.

Food is best done in a fashion that gives crew members control over what they put into their bodies. Most crew are in good shape and work long hours. Food is not entertainment for them; it is fuel to keep them

FIGURE 15.14   GIGAYACHT HEAD (BATHROOM).

FIGURE 15.15   GUEST'S AND CREW ENJOYING A COCKTAIL AFTER SERVICE
IN MARTHA'S VINEYARD.

FIGURE 15.16 THE BRIDGE (DRIVE STATION) ON A SUPERYACHT.

operating at full potential. Crew meals should be buffet style, and there should be ample food so as not to make anyone go away hungry. At each meal, I always offer a green salad with the vegetables arranged in individual piles around the lettuce so that crew members can put what they want on the salad without being locked into a tossed, dressed salad situation forcing them to pick out what they do not like and living with a dressing that they may not care for. Good quality meat or fish with sauces on the side, a starch, and a steamed vegetable would make for the bulk of the meal. You cannot just offer lasagna to a hardworking crew. Maybe leave one around for your day off, but don't lock them into meat, cheese, and pasta. Crew give up a lot of common luxuries like freedom and privacy to travel and work in this fashion. Good food is part of the benefits of working at sea at all levels. Do not think of crew food as the family meal offered to restaurant workers to get rid of leftovers. Crews eat steak and lobster, but they also may eat sandwiches and pizza. When a captain tells you that you have a budget of $20 a head per day, do not look at this as a daily budget but more of weekly budget of $140 a week per crew. One day you will make a sub sandwich bar and another day chicken, but on special days take those savings and give everyone a steamed lobster if you end up at a dock in New England. Crew really look forward to the meals, and most are on tight schedules: have crew meals up on time! If some crew members do not make it to a meal, do not take it personally, put a plate together, wrap it, and put it aside for them. Listen to what the crew wants and anticipate the condiments for them. Pick those from their countries of origin; it is a good place to start. Anything to remind them of home makes the journey less strange.

Galleys (kitchens on boats) are designed to be operated by a specific amount of people and are often tight. This can be okay if you are working alone as everything may be at arm's length and quite efficient. When it becomes tough is when the galley doubles as a passage for crew to get from one area of the boat to another. Distraction will kill a chef every time. If you have options, look for a galley that is a dead end or its own room. Often you will share it with the stewardesses who have half as a pantry. Look for two full-size sinks: one for you and one for them. These areas always look good when you are touring them, but once you fill them with workers under load waiting for the stews to hand wash the crystal and 24-karat gold-rimmed plates or do floral arrangements can add hours to your day. Another item to look for is the 3-minute dishwasher, often called the sanitizer. This is a normal-looking dish washer by companies like Electrolux or Hobart but has a superheater and can knock out many loads an hour. This can save hours of your cleanup; not only does it do the washing, but the hot pots and china also come out almost dry themselves. Read up on what can go in and always rinse everything before it goes in. Why? Because the drain is only an inch and a half, and even with a screen fats and acids collect, either blocking flow to the drain or eating away at the rubber seals. On land, this is not such a big deal, but in the middle of the ocean or in a very foreign land without a replacement part onboard you can end up with an integral part of your operation out of commission for what will feel like forever. When you prerinse in a sink, it takes the liability off the appliance and puts it in the plumbing. Plumbing is an easier fix on the boat.

The most important thing to consider when choosing between gigs is having a window in the galley. What good is traveling the world if you don't see it and take in some of that sunshine while you work? You can put up with anything for a week or a month, but try not to sign on for a season on a boat without a window in the galley; it will eat away at your soul. The rooming situation on a boat is not for everyone. We only sleep, change, and watch a movie while lying in our bunk or use the bathroom in our cabins. Most crew cabins sleep two in bunk beds and have a bathroom en suite, but if you are in a nicer boat and have a department head cabin you may get a small desk as well. Each crew member gets a small amount of hanging space in a closet and a few drawers. The trick with working on a yacht is to travel light. You will wear various uniforms 95% of the time. Have a nice pair of slacks, a dinner jacket, and a few dress shirts for going out with the boss or on a nice crew dinner or a date. For the ladies, have a little black dress and a nice outfit or two for the same purposes. We go from location to location so the locals do not know that you do not have closets of clothes to choose from; it's all new to them. A good cabin mate respects closet space, outlets, and the floor. They are clean, make their bed, and send their clothes to the ship's laundry every day. The stewardess will do loads of like clothes, fold them, and return them to your cabin so you do not have to save all your sweaty clothes for a week choking out your roommate and causing an unsanitary environment. I don't know how the stews know where everything goes: magic probably. I can tell you that I cannot account for dozens of nice articles of designer clothes over the years so it really doesn't pay to travel with a lot of fancy clothes.

A good crew cabin is kept cold and dark. It is more sanitary, and you will sleep deeper. Crew bunks are almost always fitted with thin, memory foam mattresses. I can tell you that I am a big guy, and other than size they are plenty comfortable; we usually cover up in big comforters. On the subject of snoring, get over it. Think of it as one of God's creatures in deep replenishment or purring: if you keep waking up a snorer, there will be two crew members down the next morning instead of one. If you really can't take it, and I know there are extremes, ask the captain to switch your room. Let the captain pick; don't go campaigning around the boat as you will cause secondary political problems. Crew should not be territorial, and your cabin choices should be given by rank and gender requirements set forth by the owner and the captain. You can always ask the captain about rooming with a friend, but let the captain decide. Think of yourself as more of a mariner on a crew than a little traveling family. Yes, by nature of time and location you will socialize with your crew more than most other people, but it is the yachting business, not the yachting friends.

Crew members do not wear a lot of jewelry and should not wear perfumes or colognes while working. The air is recirculated and can be spread around sections of the vessel. If the boss is bothered by the scent, they may feel uncomfortable bringing it up. Also, you do not want to compete with fragrance choices of the Mr. or Mrs. It's the same with jewelry: we are not looking for unique individuals on yachts, which is why we wear uniforms.

Yachts are built to travel, and when we leave the harbor we run 24 hours a day in shifts. The captain takes the yacht on and off the dock, but then we all share the driving responsibilities equally in the form of watch teams. This consists usually of a watch commander, a qualified or licensed seaman in charge of steering, communicating with other vessels, and adjusting navigation, and a second person, who looks out and reads a radar screen. You will be taught about the radar; it is not that hard but is very important. When the watch commander feels it is safe, they may go to the bathroom, get a coffee, or leave the bridge for a minute or two. That is why you are there: to make sure nothing happens in their brief absence and to keep each other awake after days of getting up at all hours of the day and night. Each captain will set schedule for watches to their own taste. These can be very memorable as sometimes they are up on the fly bridge on a gorgeous day, at night with a million stars out, or sometimes hunkered down in the bridge in the middle of a storm not knowing if you are going to survive. Sometimes the memories are just of the crazy conversations you will have while trying to stay awake on day 13 of the voyage. As chef, here is where it gets interesting: no matter what the sea conditions, there is someone on the bridge who is going to need to eat at some point. In the roughest of conditions, some of the crew will require food. At the interviews, then, captains do not ask if you get seasick; they ask, "Can you work while seasick?" Seasickness can happen to anyone—I have been the only guy sick out of 10 and the only guy not sick out of 17. If you are seasick, carry a gallon double-zipper zip-top bag with you to vomit into. Make sure you take the extra air out each time you add to it or it will burst

FIGURE 15.17   BRIDGE ON A GIGAYACHT.

if you drop it—not cool on a $47,000 carpet. The good thing about seasickness is that as soon as the boat stops moving the symptoms go away. Some boats actually have straps to hold the chef in place if the seas get out of control. Avoid soups and cooking pastas on those days even with your fiddles (screw-down pot racks for on top of the stove) tightly locked onto the pots. All crew will also have to assist with docking and undocking at times. You will be expected to handle lines and fenders from time to time. Docking a large boat or a small ship takes a lot of manpower, so at times it is all hands on deck.

The power structure on a yacht is a lot clearer. The *captain* is the ultimate be-all and end-all on the vessel. They trump the owner when off the dock. Their word is final, and all department heads report directly to the captain. The department heads are equal in theory, although more commercial boats have a rank order posted. The department heads are the engineer, first mate, purser, chief stewardess, and chef. The *engineer* is in charge of all of the machinery, plumbing, and electrical. They also fix the appliances around the vessel. All junior engineers report to them.

FIGURE 15.18   EVENING BARBECUE ON TOP OF A MEGAYACHT IN THE CARIBBEAN.

The *first mate* is charge of the vessel's overall maintenance and all things related to deck and safety. They are also in training to be a captain, so they aid in paperwork and navigation. The bosun is the deck manager and reports to the first mate. The deckhands report to the bosun, but in the case of the boat not having a bosun the deckhands report to the first mate. The *purser* is in charge of the cash, receipts, and crew-related paperwork. Not all boats carry a purser. The *chief steward* is in charge of all things interior, from food and beverage service to cleaning laundry. They are the host and chief housekeeper. This is a varied and difficult job carried out over long hours with grace in front of the boss. Don't give them any shit; their job is harder than yours, and they are not there to serve you. All junior stewards, laundry staff, and spa workers report to them. Ideally, the chef should have the final say on all things related to food and service, but good luck with that as long as southern hemisphere (Australia, New Zealand and South Africa) and British women hold the vast majority of these positions. Try to work as a partner and don't make eye contact; they can smell fear.

The *chef* is in charge of catering and all things galley: menu, food, and food storage as well as the cleaning and maintenance of all galley-related appliances. The sous chef, crew chef, pastry chef, and utility and scullery help report to the chef. Who are these rock star chefs circling the globe?

They are male and female; and as I have stated, this industry for many reasons has a high percentage of successful female chefs. These chefs are young (thirties are the peak years), single, experienced, trained professionals who are presentable and outgoing. Yachting is the third most image-driven business on earth after modeling and acting. You have to be healthy-looking, not gorgeous (even if you aren't a gym rat). You must look good in your uniform since you will inevitably be invited to the table of one of these floating jewel boxes to be shown off by the boss. A good chef is a huge source of pride among the wealthy. These chefs are from every corner of the world. They are largely nonsmokers and can pass background checks and drug tests. They have valid driver's licenses and passports. They all speak English, even if it is not their first language, but they can change languages, currencies, and types of transportation within the same day as they island hop internationally. Some have never worked in a restaurant, and some have earned Michelin stars. They are self-starters, can work unsupervised, and are healthy enough to be on their feet 18 hours a day on a moving platform. They occasionally get seasick but can work anyway. They are clean and can live 24/7 with crew from all over the world without starting fights. They can sleep in a broom closet with an exhausted, snoring roommate without waking them or stressing out over it. They all have taken a five-day safety course called an STCW-95 that covers firefighting, sea survival, first aid, and social responsibility at sea. They are savvy at networking and know when to charm or avoid someone. They respect authority and can work creatively within structure. They have a sense of humor and know when a girl needs chocolate. They are good listeners and detectives. They do not say no to the boss unless the request is impossible. They communicate efficiently and politely with the service staff. They must be good dancers and should know how to juggle to some extent. They are resourceful and usually know where to get the best of everything and where the coolest crew bars are within a few hours of arriving in port.

Does this sound like you? Oprah Winfrey once said, "Luck is when preparation meets opportunity." If this is you, and if yachting is in your future, the odds of you having someone come to your village and offer you a superyacht gig are statistically slim. You need to go to where they do business; it is not good enough to see one in Nantucket or Capri and tap on the hull. They are operating full crew at that time, and if someone lets them down they are most likely going to go with a proven next candidate, not a stranger newbie. As with love, if you are going to become a yacht chef you are going to need to enter with reckless abandon or not at all. You are going to need to go and establish yourself in the world headquarters of yachting: Fort Lauderdale, Florida. All yachts are bought, sold, crewed, and fixed in Fort Lauderdale. It is the Venice of America and has the facilities and craftsmen to fix and refit them. There are miles of marinas, shipyards, and warehouse districts that stock and sell every conceivable resource a yacht could need, and they all deliver. The 17th Street causeway is the center of the universe for yachts; at its peak, a small section between the Burger King and the bridge did billions of dollars in business discretely from small, unassuming offices. More than 50 crew houses are on the back and side streets to house the transient crew in between jobs or studying in one of the maritime schools in the neighborhood. There are yacht crew agents lining the streets and resource centers for getting your paperwork in shape and competitive. Prices on everything in Fort Lauderdale are cheaper than other ports we visit, and the nightlife and entertainment there is on a level that leaves both owner and crew satisfied. This city is to the yachting business what Hollywood is to the movie business. I have maneuvered both, and they are shockingly similar, with Hollywood being a little more mature in its process and taking its offerings to the public and Fort Lauderdale taking its goods and services private. Both cities have fiendish spending and epic parties. Beautiful woman, unlimited booze, drugs, flashy cars, and every conceivable distraction abound. There is simply a ton of money being spent by shiny, relatively uneducated people there. Many yacht crew get sucked into the local party life and never make it out of port. In the summer, Newport, Rhode Island, becomes the satellite capital for yachting in the United States. It is like our summer camp—and a camp is fun! Antibes and Monaco, France and Palma, Mallorca are the European yacht capitals for the Mediterranean summer season. St. Martin is the satellite capital of yachting for the winter season. These locations are all seasonal and deal with provisioning, additional crew, and repairs while in season; however, all these satellite locations revolve around Fort Lauderdale, where we go when we are not in full service.

Smaller to mid-sized yachts will have one chef who is charge of all things food and beverage and works in concert with the chief steward. This includes planning, provisioning, and all aspects of cooking for the crew and guests, cleaning all pots and pans, and detailing the galley. The stewards clean all of the dishes and glassware. Cleanliness and organization are the hallmark of a good yacht chef. At 52 meters (170 feet) you may get occasional assistance from a junior steward to help count inventory, organize stores, or detail around the galley. When

FIGURE 15.19   NOT ALL DAYS ARE FUN OR EVEN SAFE AT SEA; I WATCHED THIS HAPPEN AND GOT SOAKED FROM THE WAVE IT CREATED. IT WAS FRIGHTENING.

FIGURE 15.20 YACHTS IN ST. MARTIN, HOME PORT TO MOST CREW IN THE WINTER.

you get closer to 60 meters (198 feet), you start getting help, usually in the form of a sous chef. This can really make a difference in your quality of life. The sous chef is usually someone who wants to learn from a master of their craft or is on the road to being a solo yacht chef and wants to reduce the learning curve. They can do crew food and prep, you can trade off meals; they do breakfast, you do lunch, you both do dinner. This allows for the head chef to fly out for the weekend and leave the sous chef a chance with their own galley for a while; it's a win–win!

When it comes to a gigayacht, it can be fantastic for a head chef. I have done two with a crew of 30 or more on each. The galley has a full staff of six: head chef, sous chef, crew chef, pastry chef, and two utility guys who wash pots, pans, and dishes as well as detail and fetch well-hidden items from the stores below. They also will peel and chop and DJ if you spoil them a little. You have a whole team to share the cleaning and organizing with. To a good leader, it is like conducting an orchestra and can be a gorgeous experience. If you are stepping out of your league, it will be very stressful on everyone onboard. Make sure you have the self-esteem and self-confidence to accept a gigayacht executive chef position: this is not for amateurs.

FIGURE 15.21   THE FIRST OF MICROSOFT CO-FOUNDER PAUL ALLEN'S FAMILY OF
SUPERYACHTS, "MEDUSE." MR. ALLEN WOULD GO ON TO PIONEER THE GIGA-YACHT MOVE-
MENT, BUILDING A SUCCESSION OF LARGER AND LARGER GIGA-YACHTS, EARNING HIM AT ONE
TIME THE TITLE "KING OF YACHTING." THIS 200′ER REALLY TURNED HEADS BACK IN 1996.
LATER BOATS IN HIS FLEET WOULD TOP OUT AT OVER 400′ IN LENGTH.

At the 100–140-foot range, most galleys are like home kitchens with fancy name brands and underpowered equipment that looks familiar to the Mrs. in an effort to aid in the sale of the vessel. When you hit the 150 feet and up, things start getting more professional and eventually downright commercial. You may turn the flattops on a gigayacht from a panel on the wall 6 feet away. Open flames are frowned upon by the insurance companies, so it is mostly electric and induction cooking. Gas grills can be on deck but are useless on a windy day, and gas types and connectors vary around the world. However, when they are usable they can really make the meal. A smaller boat will have a home refrigerator in the galley, a chest freezer in the lazarette, and storage in cupboards and under the seats of a banquette. A 50-meter yacht (164 feet) will have a walk-in fridge and freezer in the bilge or crew quarters and a pantry closet. A gigayacht will have up to five walk-in refrigerators, a climate-controlled pantry room, and a produce or fruit fridge.

Don't expect your own room at any level, but occasionally you get one: be grateful. Your food and travel are paid for. Most mariners are considered under the care and cure of the vessel. They supply everything from toothbrushes to razors. If you are hurt on the boat, they cover it. Most crew get their own health insurance and then have the boat reimburse them for it in cash. It is better for both the crew members and the owner for it to be that way. Your insurance comes before the boats, and you can take it with you from boat to boat with no gaps. In terms of travel, the boat arranges the tickets and pays for ground transportation from the time you land onward. You usually pay for yourself to get to and from your home airport. The boat should pay for any baggage fees, and some programs pay for meals when traveling; therefore, save receipts from everything. Receipts are valuable in the yachting business: we buy them with cash. If you are on smaller boats, use soft luggage (duffle bag with wheels); on bigger boats, get yourself matching, rigid, lightweight four-wheel spinner luggage (one carry-on and a full-size). Do not buy designer, but also don't buy cheap. Just trust me on all of this. Get to the airport early because rushing into a new program is stressful. Arrive early, get something to eat, and keep hydrated. Dress nicely while traveling, and people will give you less hassle. Always have cash and a debit card for in-flight purchases, and just before boarding change all your currency to that of the country to which you are heading. This will ensure you will not have to hunt down an exchange at your destination and will be able to hail taxis and buy snacks immediately upon arrival. Yachties (private yacht workers) travel more than anyone I know. We live and work on a moving object. Get good at traveling.

On yachts you have responsibilities that are not at all related to your land-based work. You are a mariner with safety and security responsibilities. We work 24/7 and are paid that way. When the boats travel, we store all movable objects, tie everything down, and run 24 hours a day for thousands of miles at a time. The owners will usually fly in and meet the boat at its new location when there are major changes of venue. You may think that they buy a large yacht and ride across the ocean on it: not the case. Hard to believe while in port at rest that the boats can rock so hard at sea in a major storm that a wall can swing up and knock you unconscious. I have been on boats that have taken on serious water in storms far offshore and almost died at sea twice. I have known several people who have had the ship go down, most saying that one minute they were in bed and within two minutes they were floating in a survival raft alone in the sea waiting for help. I have known people who have died doing this. We do amazing things with amazing pieces of equipment: we cross oceans and dance with Mother Nature. James Bond would not be James Bond without the danger.

You may notice that I didn't talk much about food. It's a small part of the job in a 24-hour day. It is the best part though: cooking Italian in Italy one night and French in the South of France the next, all using local vendors and markets. Learning the local languages and immersing yourself in the culture and history will make you a much better chef. If you truly want to know French cuisine, go to France and eat at the local bistros, drink the local wines with the vintners, try to rent a Peugeot using no English, and take on a temperamental French lover for the duration your stay. Cooking is great, but I did a lot more listening than talking while I was there. We never really talked about cooking. We did discuss the ingredients of course, but the dishes just sort of happened three times a day everywhere I went. Italy made me understand the pace of a meal and that the ingredients had to be the star. The Italian chef was a solid custodian of both freshness and tradition. I paid a lot more attention to the food in Italy because the women were much harder to make your lover. All French girls spoke some English, but only 1 in 20 Italians spoke any. I was a more of a silver-tongued devil: if I couldn't chat up the ladies, I would at least eat well. These things you can know only from being there. I could not develop true passion for these cuisines from the confines of a restaurant office and a stack of books. Trouble is, the only way to go circling the globe with your own kitchen—not only cooking but also eating and drinking in the culture—is to get on one of these boats and go on a global adventure thanks to your billionaire employer. As they say, "It is good work if you can get it," and nothing will make you a better chef than this type of gig. Nothing.

CHAPTER 16

The Islands

Working and living on an island is truly a unique experience. The remoteness, even if only by a moderate ferry ride such as the journey to Martha's Vineyard or Nantucket, causes a stronger sense of community, which can be good simply because people look out for each other and pool resources when the going gets tough. Those who live on islands seem to take a keener interest in fellow islander's affairs, but this can be challenging for those with high-profile clients who are trying to maintain their privacy. For all the inspiration being surrounded by water brings to people's lives, it also creates additional challenges to basic necessities that those who live on the mainland take for granted.

Schedule limitations of ferries and small planes can wreak havoc on a special occasion that counts on last-minute special orders, seafood, and produce deliveries. Rough weather can put a halt to all traffic on and off an island. Local power grids are also a concern; the small generating plants that take care of most islands are far more susceptible to outages than any country's national grid, making on-site generators no longer a luxury but a necessity. A moderate storm can easily knock out power for twice the amount of time that people face on a nearby shore.

International islands that are far offshore; such as the Caribbean, Seychelles, or Polynesia, have wonderful lifestyles and lots of natural

FIGURE 16.1   NANTUCKET ISLAND IS FREQUENTLY CITED
AS MANY YACHT CHEFS' FAVORITE PORT STOP.

beauty. They are easy to visit and vacation on, but living and operating a culinary program in these locations can offer many challenges for both public and private chefs.

Food and dining are taken more seriously in vacation destinations, so the availability of quality standard ingredients is not usually a problem on larger islands; however, with the longer travel times foods often arrive toward the end of their shelf life. If used within two days you are fine, but if you are shopping for a week out do not count on perishables making it. When you need something that the island does not stock, the item must be brought in by special order and the shipping, and method of travel to arrive to the island is a challenge over which you have no control. Most frustrating is when you get confirmation that your shipment may be late but did arrive on island earlier in the day and no one can locate it. Corruption and low accountability are ways of life in a banana republic. If you are lucky enough to dodge that bullet, there is also customs to pass and sometimes tariffs to pay. Once it has gotten that far, you are treated to the local delivery system where so much can go wrong.

On smaller islands of under 100,000 people, another crisis can occur. If the Heineken boat does not show up on schedule, you would be surprised how valuable a case of Dutch beer can become. People will

FIGURE 16.2   THE VIEW FROM A SUPERYACHT AS YOU PULL INTO SIMPSON BAY, ST. MARTIN, THROUGH THE BRIDGE. BEST WELCOME EVER.

FIGURE 16.3   THE MAIN DOCK AT DAVID COPPERFIELD'S MUSHA CAY IN THE EXUMAS BAHAMAS. MOST CONSIDER THIS THE CROWN JEWEL OF PRIVATE ISLAND EXPERIENCES. THIS IS THE ABSOLUTE MOST PRIVATE GETAWAY POSSIBLE FOR THE LIKES OF OPRAH WINFREY AND THE ROCK BAND U2.

FIGURE 16.4   NEWPORT, RHODE ISLAND, IS MY SUMMER AND FALL HOME PORT FOR A
REASON. IT IS NOT ONLY THE SUMMER YACHTING CAPITAL OF AMERICA, BUT IT IS ALSO
THE SAILING CAPITAL OF THE WORLD AND SATISFIES BOTH OWNER AND CREW WITH ITS
CHARM, SOPHISTICATION, AND OVERALL ATTITUDE OF CELEBRATION.

hoard their supplies, and without well-cultivated friendships borrow-
ing some to get you through the next few days may become just a
pipe dream. Vices are cheap on duty-free islands. It is always a good
idea to invest in some goodwill with local chefs; buy a few rounds for
them from time to time or perhaps patronize their restaurant and be
complimentary. When things get tough, it is a much better feeling to
have a friend return a favor than to try to negotiate a few loaner cases
out of desperation.

Living on an island, it is fun to hear about the other residents' business
and affairs. Everyone seems to have a great story about how they ended
up there. It's OK to share your own tale and make lifelong friends and
allies, but be careful to keep it separate from the boss's. In the long run
you want to have people get to know you and your story while adding
to it, not to be looked at as a reporter from behind the gates of the boss's
property. This can be difficult since so much of the life of a private
chef is spent doing work-related activities. If you want to be successful
in a microcosm with jealous townies with low self-esteem trying to
overhear anything salacious or novel, keep your conversations focused
on the situation at hand. Talk about what you are doing at the present:

FIGURE 16.5 ATLANTIS, NASSAU, BAHAMAS, IS ONE OF THE WORLD'S PREMIER MARINAS
AND A GREAT ISLAND EXPERIENCE.

the drinks, the beach, and the new band coming to the local pub next weekend. Make plans to go sailing with your new friends on your day off and leave the work at work.

Real police are less of a presence in this world. Many islands seem to have their own system of justice; islanders take care of their own business, and law enforcement officers seem to investigate and clean up later. The only place I have ever seen people dead in the streets were on islands, but the good news is that for every dead body I saw I witnessed a hundred live people fall in love. Take it from a man who lived and worked on many islands for most of his adult life worldwide and never had a safety issue, if you want to avoid trouble on an island then be where you are supposed to be. Travel in groups when possible. Most trouble comes when someone ventures off of the beaten path trying to score drugs or find prostitutes. In my experience, most island crimes happen between midnight and 8:00 a.m. In the Caribbean, if you are not home in bed at midnight after a long day of fun in the sun you are wasting your time and money. I promise you, the drinks and the sex won't get any better or cheaper after midnight. Get some sleep and save your money.

Figure 16.6   Weekend Harley rental in St. Martin.

PRIVATE ISLANDS

There is no doubt that being in a major city has its advantages: unlimited shopping, entertainment, and social opportunities. Anything you need is right there, and there are lots of people around to play with. Big cities have all the stimulation you can handle, but it all comes with a price: crime, rules, regulations, traffic, and huge limitations on privacy and personal space. I had one of my most amazing experiences doing a gig on one of the world's most elite private islands. It wasn't for what it had—I don't think there was anything there physically that you could not find in other places—but more for what it did not have. It took quite a journey to get there: commercial flights, a limo ride, and then onto a private jet boat that kept us at sea for nearly an hour at high speed. The journey was long enough to give you time to realize that this was a very special place with no close neighbors and true privacy. There were no municipal authorities or any government on island—basically no one to tell us what to do. There was a small group of department heads (chef included), and we each received our own staff cottage and golf cart. Three times as many local island understaff stayed on one end of the island in a makeshift village of dorms, a cafeteria, some workshops, a commercial laundry, and a shared pool of golf carts. Many of them would come in for charters or when the owner was in residence but then returned home to the big island. The island had a series of cays surrounding it, one with its own airstrip and a couple of turbo prop airplanes that came with the charter. The main island had a heliport and a fleet of various-sized boats and jet skis. It featured a series of gorgeous homes and common buildings. Every inch of this island was meticulously planned, executed, and maintained. It had an almost Disney feel to it. Electric golf carts were the only form of land transportation. It was all fabulous, but the cool part about it was that no currency was used— no cash registers, no stores. You had the best of everything in abundance, and all the vehicles were just there, all gassed up with keys left in them. No crime, no prying eyes, or paparazzi. One A-list celebrity and his Playboy playmate girlfriend wanted to roam naked during the afternoons, so the managers made an announcement that no one could be on that end of the island during those hours. If there were security or paparazzi concerns, the staff could ask the nearby government to host a no-fly zone around the area, and the local coast guard would do the same for boats. This was something most people will never experience; it was very freeing. Staff was rarely seen until needed, and certain people were there to take orders and disperse personnel and resources wherever necessary. As staff, we were living far out to sea with little chance of leaving the island. We were highly trained professionals and deserved some perks as the owner raked in over $1 million a charter. After dinner service, you could raid any bar on the property as long as you were not seen. You could take a bottle of anything you wanted back to your cottage and fill your bathtub with Häagen-Dazs if you so pleased. Epic karaoke nights followed five-star staff dinners, and we could take the jet skis anytime we wanted as long as no one was in residence. There is a different mentality to watersports when you know no one will come to save you if things go wrong.

Obviously, I am talking about one of the most elite private islands; I just want you to understand that this level of opulence does exist and chefs are needed. The principle is still the same on private islands everywhere. The beauty, remoteness, and lack of governance comprise an attractive situation to many, and the world's private islands that

FIGURE 16.7   SIMPSON BAY, ST. MARTIN.

FIGURE 16.8   THE VERY PRIVATE CAT CAY IN THE BAHAMAS.

have been sold or developed in the last 20 years far outnumber the megayachts and superyachts out there.

When operating a culinary program in this situation, you need to be highly organized and fastidious. If you do not order correctly and stock up on inventory, it can be very difficult to get more stock and expensive if you have to send a private boat or plane long distances for a few items—not to mention during that time you lose on island manpower.

No one wants to be the chef who served $3,000 peanut butter sandwiches because the boss's kids wanted them and you forgot, so you then had to send the pilot to the mainland to pick up the peanut butter.

In addition, if a fridge or freezer goes down, not only do you lose the value of the food, but also you no longer have food to sustain the inhabitants. You must monitor equipment regularly and learn to fix your own gear with a minimum of jury-rigging. Insects, clever animals, and improperly wrapped or stored food can be a recipe for disaster in some of these locations. Inventory management counts so much more. I also would not recommend working on a private island without any first aid training.

If you only travel to these locations on occasion, they are novel and fun, but if you are going to spend any amount of time on a private island you better bring a lover, because there are only the staff and the boss's party to socialize with. It is rarely a good idea to mix with the boss's party; there are exceptions to this rule, but do not count on it. And mixing with other staff in a casual way is not a good idea as it can create an uncomfortable work environment for your fellow staff and create drama on the boss's dime and time. If you truly fall in love with each other, go for it, but if it's just a fling it's best to avoid it. Couples do most island jobs because it keeps the staff from roaming and overmixing with the locals. To cut down on cabin fever, a good computer will help, and most billionaire assets have decent, if not state-of-the-art, communications. You will come to count on this to send order lists and to live vicariously through social media.

CHAPTER 17

The Planes

Nothing has changed the private service industry, as a whole, so much as the advances made in technology, model options, and infrastructure of private aviation. The wealthy, with their newfound superfortunes and instant global communications, can now be anywhere in the world on a whim or to stop a corporate meltdown within hours—no reservations needed. Business jets have been around since the 1960s, but most are known as *light jets*. These are small, private, jet-powered planes that are more like limos in the sky. They are designed for privacy, security, and efficiency in regional travel. They bypass all of the security and boarding procedures that the masses face at commercial airports. The boss pulls right up to the tarmac in their own limousine, steps onto the plane, and takes a seat; the pilots and limo driver load the bags; the copilot hands the boss a beverage; and off they go. Pilots will often grab a deli platter for self-service on these short flights. There is no kitchen, flight attendant, or prep area. The boss has a stereo and TV but cannot stand up; if they're lucky there is a portable toilet in the back. This was considered enough luxury as these planes were used exclusively as transportation.

There are about 500 airport hubs that can handle a commercial jet airliner. There are more like 5,000 regional airports that can handle the needs of a light jet as they take off and land in much shorter distances. A light jet usually holds an average of four to eight passengers. Private jets fly higher and faster than commercial jets, usually above the weather, which makes the ride smoother and more fuel efficient. On domestic flights, a limo will be waiting on the tarmac to whisk the boss to and from a luncheon or meeting before reboarding and heading for the next stop. Four cities in an afternoon, no problem; face to face to stop a crisis in real time, priceless.

With the average light jet having a service life of over 30 years, they pay for themselves in time with corporate productivity and as perks for executives. For many years, the corporate jet was just that: corporate. It was looked upon as a status symbol to own or have use of one, not because of its luxury appointments but more because it sent a message of importance. Good examples of a light jet would be a Lear Jet 25 or a Hawker H-400. Kings, presidents, and sultans have converted corporate jetliners for decades. All this started to change in the late 1990s: technology and communications made leaps and bounds, and instant millionaires and billionaires realized that for a while anything seemed

FIGURE 17.1   TYPICAL BUSINESS/LIGHT JET.

possible as rules and regulations had not been written or implemented for their newfound situations. Mobility was prioritized, and nothing allowed this so much as your own private, not corporate, jet aircraft.

The mid-sized jet market exploded; orders for more comfortable, longer-range aircraft poured in. Now your own plane could cross oceans without refueling. These planes had custom handmade interior, featured satellite communications, and even sometimes enough headroom to walk through upright. You could hold a meeting, catch a nap, or watch the game. Most of all, these jets now featured the ultimate luxury: staff and a galley. These mini-jetliners seated around 8 to 12 passengers (each seating plan was custom), two pilots, and a steward. Families and corporations now could expect gourmet service at the level of commercial first class or better. That being said, no cooking occurs on a mid-sized jet. Power restraints, cabin pressure, and venting all restrict this, but an induction burner and electric oven/microwave can work, even though they are strictly for reheating foods that have been prepped and cooked elsewhere. How does this affect you? You are now their sky chef. Yacht, estate, and corporate chefs make all preparations on the ground, seal the foods in airtight containers, and send them in soft cooler bags to be assembled by the steward on the plane. This sounds easy enough, but there are some things you need to know. The boss does not eat leftovers; you do not fully cook anything but a fish or poultry. The problem is that there is no guarantee that the steward is going to know how to read the food for the finished cooking and plating as many work only freelance. Your billionaire may have a solid staff pool for security reasons, but on average most are called in on short notice. Communication is also important, as you may not want to pack a chicken supper if the boss had chicken for lunch; therefore, it is best to coordinate menus in advance. Another tricky element to factor in is that at certain altitudes and with cabin pressure at play you lose 10% or more of the effectiveness of your taste buds. In private service, the plane does not always leave when you expect, but you have to anticipate when the dining will take place. If you know that the plane is to travel on a long uninterrupted journey, you may want to season heavier. If the boss's kids get hungry or there is a delay and they decide to pig out while they are watching a movie when waiting on the tarmac, it can be a disaster. If you season normally and they leave on time to settle into a nice dinner halfway to Paris, you may be serving bland food. No one said being a private chef is easy, but good communication can trump good cooking most of the time in the pursuit of success. Make sure that you include a menu, a drawing, or picture of the final plating; mark containers clearly with instructions; and double—no triple—check everything before it leaves your care. Choose garnishes ahead, and pack so everything has its best chance of survival. Include your business card in the package so the steward can call you if there are any questions; don't assume it is your regular steward.

The chef will normally arrive a day or two ahead of the boss's party to get the site ready, but although this scenario is rare they may occasionally fly with the boss. Keep your cool on the plane and let the boss do the talking. Often the conversations you have will be ones you will remember for life. Even though it may not be that sophisticated, the guy who invented rubber dog puke has a success story.

It would be in your best interest to visit some of these planes and see how small the galleys are; everything on a plane is about being lightweight, taking up no extra space, and being ultraefficient. The planes have their own custom plates and china and very little counter space.

FIGURE 17.2 SOMETIMES EVEN THE COOK GETS A LIFT.

Examples of a mid-sized private jet are the Gulfstream G-4 or Bombardier Challenger (which are considered large jets in the private-jet catalogue, but are considered mid-sized in the billionaire scheme of things). These are the ones your billionaire is most likely to own or participate in a fractional jet ownership pool with. Most billionaires will own several aircraft so that they do not get stuck using a giant aircraft for regional hops or have to fly commercial on long trips due to lighter aircraft range restrictions.

If you are working for one of the ultimate superwealthy or for a sports team owner, there is another kind of private jet you will encounter: the private jetliner. These are large commercial jets that are factory configured for large groups and entourages with spectacular range and capabilities, the most common being the BBJ (Boeing Business Jet). This is a 737-700 series featuring an 80-foot cabin and can have bedrooms, a dining room, office, and a main salon (living room). Sports teams will have a full staff and 30 first-class seats, each with their own entertainment system. Occasionally, they will be outfitted with a galley and feature an in-flight chef. The galley will be similarly outfitted for a small condo: four burner stove, oven, warming drawers, refrigerator, cappuccino machine, and microwave.

The ultimate experience is working for a head of state, sultan, or sheik. They will order jumbo superliners with multilevel floor plans. These planes, such as the U.S. president's Air Force One, are a Boeing 747 featuring two full galleys for both convenience and security. Redundancy is a big factor in aviation safety, but in the case of a national disaster Air Force One may have to stay airborne for days. These planes can refuel in flight and remain airborne indefinitely, and as long as one galley is operational they do not have to land. I once spent an afternoon in St. Martin with two American stewards working for an oil sheik who owned a brand-new Boeing 777 jumbo liner with a nightclub on the second floor. The staff stays in nice hotels with ample per diem wherever the plane is when the boss is traveling. This can be a cool job except that these planes do not fly as high as the purpose-built corporate jets, which fly higher than the weather. Cooking in turbulence cannot be fun. To address this problem a few of the corporate jet makers such as Gulfstream and Bombardier have produced ultra–long-range, longer versions of their mid-sized jets, which fly anywhere in the world with a maximum of one fuel stop. They, too, fly high above the weather, and although they come with narrower interiors the trade-off is in-flight chef capabilities. These are provisioned in a similar fashion to the regular mid-sized corporate jets.

The fact that your boss has a flying dining room or kitchen is not what has affected the private service industry so much: now, the boss can be anywhere in the world without any hassle and without losing any efficiency or comfort level while in the air. The boss can now decide to blow off the rest of the day while at lunch in Manhattan, have their spouse meet them Teterboro Airport across the Hudson River in New Jersey and have dinner at their home in Paris or St. Barths. It is not that you will rack up private frequent flier miles—those jobs are an anomaly and not an industry—but more chefs have to be on retainer or employed in more locations to make all of this happen. Kudos to the private jet industry.

CHAPTER 18

Cooking for Celebrities

Personally, I am and always have been drawn to all things celebrity. If I hear a celebrity is in the neighborhood, there is no way I could not somehow change my behavior to up my odds of making a sighting or perhaps an introduction. I do not want or need anything from them; I just think that popularity is interesting and fun, and apparently I am not alone. I have worked for some of the wealthiest people on the planet, and sometimes our trips with them are very exciting. When I come home and try to tell a story about those experiences, though, people's eyes soon glaze over. However, if I come home after having had a boring trip with a celebrity and their family vacation, I not only have the listener's full attention but they also grill me on every detail.

I had met some major celebrities while working in fine hotels and when I was in the music business. I had had brief conversations with Neil Diamond and Carol King alone in their hotel room or dressing room, but the first time I ever really stepped all the way into that world and cooked for one of them in a private setting was for Calvin Klein at a house party on an island in Miami. I was asked to fill in for another chef who was double-booked. Star-struck for a few minutes, I soon realized it was time to go to work. The work did not change. No one hired me because I was a friend to celebrities or because they thought the celebrity would think I was a cool guy to hang out with. They hired me because I was a focused, talented chef who could handle the job at hand.

Food styles and quality will change from client to client, but celebrities do have some unique requirements. Billionaires may have scheming bad guys after them for their money or for some sort of revenge; these are straight-up security issues. There are well-studied situations with protocols in place to prevent and protect them. Anonymity is the best offense with the ultrawealthy. I can walk the streets of the Hamptons or Nantucket with many of them, and almost nobody knows who they are. When we all walk off a yacht to have a group dinner, most people have no idea who the boss is because we all dress nice and walk in a tight group.

On one occasion, I had a gorgeous female client in Miami who was from Mexico. I would take her 12-year-old son fishing. Once, I asked him what the difference between Mexico City and Miami was. He said, "Mexico City is not secure": a pretty heavy response for a preteen. In Miami his mom would ride around relaxed in the front seat of my car (I personally owned a decent car) in a Juicy T-shirt, skinny jeans, and sunglasses. We drew no attention to ourselves, and she could go about her business unencumbered. Even I had no idea that she was worth over $2 billion and was Mexico's richest woman. She requires a motorcade back home.

Celebrities do not have this luxury. Stalkers do not often know where a celebrity is vacationing, so they are not a major concern for us. Where we do see problems is when they are noticed while we are out shopping or dining with them. Sometimes you get an impromptu mob. If you are on an island or on a yacht, people sometimes find out where they are staying. Some become curious or even temporarily obsessed, which can complicate your situation and be downright dangerous.

Oftentimes the wealthy will find it novel to have celebrities travel with them. This can be a lot to handle as in these situations the boss may like to show off their new trophy friend and enjoy the newfound attention out on the streets. Making this worse is when the boss invites half the bar's patrons back home at closing time to create some cool memories. We have to make sure that everyone is safe and that nobody is crashing totally uninvited.

FIGURE 18.1   CHEF NEAL AFTER TRAVELING ON A YACHT WITH DIDDY.

I used to do some work with Sean Combs—Diddy to the masses and Mr. Combs to me. When a star is a wealthy and famous CEO, security is escalated. At this level the client has a bodyguard who takes care of them and pretty much only them. If the bodyguard escapes a bad situation with the client and we all perish, he still has a job—and probably a raise. When we are having many guests, outside security is brought in. Usually, local police detectives and some uniformed officers are used. This takes the liability off the bodyguard and us and puts it right on the police. Mr. Combs chartered a yacht in a major U.S. city to host a several-day event. Most law enforcement officers are not rich. Most don't hang out on yachts. Most are not savvy elite party pros. This became painfully apparent on one occasion when Mr. Combs was off the boat for a few hours to make a public appearance. A chubby detective, in an effort to remove his shoes, bent over in front of the automatic sliding glass doors on the aft deck. The doors shot open just as the mace on his belt was squished and went off. Boats have semiclosed air handling systems, so the mace got sucked into the vents and the whole boat—thousands of square feet—got maced!

While I was waiting for the air to clear inside, I walked to the foredeck, where two middle-aged, youthfully dressed ladies were sitting and chatting. All of our guests were off the boat at the event. I asked who they were visiting, and they dismissed me, saying not to trouble myself as they were friends of the owner. The boat was being chartered, so it was possible that they did know the owner and were unaware that he was not in residence. I interviewed them, and they did not know who the owner was or even who was chartering it. They had walked right passed the police detail, who probably helped them get aboard safely. They would not leave. I made one radio call to the bridge, and they were escorted off the boat in handcuffs. We opted not to press charges, and they were released outside of the marina gates. Security comes in layers, and we all have a role in it.

Another layer of security you play a part in is keeping your mouth shut when out provisioning or socializing. There is a great temptation to tell people you are traveling with an icon. You can become the star of the bar—maybe even get some drinks comped. This puts your client and your team at physical risk. You never know who is listening or who they know. We are highly trusted and paid well to protect our clients, so do not undermine that trust. Celebrities and paparazzi have no problem sharing details of a vacation after everyone is safe at home. Afterward, tell your stories, but do it like I'm doing here: nothing about the private details of their personal business. You could hold a check to my head, and you won't get dirt from me.

In 2001, Whitbread around-the-world sailboat race winner Sir Peter Blake was killed at a stop on the Amazon River. Local criminals noticed the crew members on Blake's *Exploration* yacht wearing matching luxury watches, rumored to be gifts from him. The locals talked with the crew about their trip. The criminals in the crowd probably even bought them a round before later assembling a team to rob them at gunpoint. Blake was highly outgunned and killed trying to defend his crew. It was not the crew's fault. We all want to bond with the locals on our trips. Making new friends is part of the joy of travel; it is human nature. The sad part is that there are criminals who lie in wait until an opportunity falls in their lap and awakens them. Traveling can be lonely, and there is a great temptation to share your world along the way. Be careful.

The next consideration is the working nature of celebrities. I have done some work with Martha Stewart, another media tycoon who makes her money on both sides of the camera. Although she has the means to own a yacht, she prefers to be known in the yachting world as a "good guest." On several occasions, she has popped into my galleys totally unannounced. I am always glad to see her, and if you really have your act together she genuinely respects it and can be very flattering. I have yet to disappoint her. I don't think that would be fun.

When I see top-tier celebrities chartering a large yacht, a monster villa, or chateau, it is usually when they have just finished a major project and are on top of the world or the charts. Sometimes the studio or the parent company picks up the bill. It is fun to see them dominate the TV in the crew area while they are upstairs having a nice, relaxing vacation. When someone has given every possible piece of themselves to their art and only has a short time to recharge, I try to take their time seriously and not give them any reason to stress.

Celebrities who work a lot end up eating a lot of craft services and restaurant food. That is not to say that it is crap food, but when we are not selling food we do not have to worry about markup. We have access to foodstuffs that most celebrities have never encountered. We can blow their minds with far fewer ingredients. From assistants to managers to agents, people always surround celebrities. In public, the whole world is watching. Most celebrities do not have a live-in, full-time chef as few proper meals are taken at home. Many have a cook or housekeeper who occasionally does some meal prep.

Can you imagine? After all the public drama that comes with stardom, to have ultimate privacy, with a world-class chef and a discrete service team dedicated only to their needs—with only the people they choose, love, or want around them. Nobody knows they are there. This is a dream for a public figure. There is a reason that people pay us so much for this.

There are a few other considerations when working with the famous. Sometimes you will work for celebrities while they are on the job. At these times, you know that they are under pressure and really need to have their back. Do your homework, stay on top of things, and be absolutely ready for anything on a moment's notice. Never make them wait or worry.

I did an event with Martha Stewart a few years back in New York City. It was held on a longtime, on-and-off-again client's yacht who would be involved with the event. Media from all over the world would be there. I had about six weeks' notice, so I had plenty of time to get things organized in Manhattan while I was still in Florida. Sherlock Holmes couldn't have been a better detective. I used the meat shop where she wrote the foreword for their cookbook. I found the bakery that does the pastries for her show. I looked online, and I talked with my mother, who is a big fan. I hired a farmer's wife, whom I had met in St. Martin and is the self-proclaimed biggest Martha fan on earth to be my kitchen assistant for the week. She went straight from a potato farm in western Massachusetts to having a stateroom on a 140-foot Feadship (the Rolls Royce of motor yachts). She was a great resource on all things Martha, and she even got to meet her idol. I had left nothing to chance. Everywhere Martha looked, there was something familiar and to her standards. Her team had been there for a week in advance decorating,

Figure 18.2   Chef Neal with Martha Stewart on a yacht in Manhattan.

and we had the catering covered. The event went on for two straight days: many meals, meetings, and broadcasts. Martha was there for only one day, but everything we did represented her standards the whole time. She was very appreciative of our efforts. With my client's wishes firmly grasped, my galley was calmer and more organized than it ever has been. The farmer's wife was an excellent choice for an assistant—lovely and dedicated to the task.

There is one other consideration when working with the famous. While sometimes they mix business and pleasure, the famous tend to be pack animals. When you are traveling with a famous mogul like Diddy, they are rarely traveling alone. Often they will have a mix of family, friends, fellow artists, and executives related to their business. This can be fun when you are asked to make and deliver snacks to the week's chart-topping hip-hop artist on the fly bridge and then run down to bartend for the world's biggest boy band and their entourage in the main salon. Where it gets challenging is when all of these people are on completely different schedules. The mother-in-law may not be on the same schedule as the performers and executives who make their living in hip-hop and expect their main meal at 4:00 a.m. after coming home fresh from the clubs.

I remember on one trip in particular the yacht was starting to feel like a hotel. Someone would call into the galley and ask for a chicken sandwich (for some reason we had only chicken on the bone left so deboning was done to order), hand-cut fries, and so forth—and when it was being delivered someone would see it and say, "Hey, that looks good; I want one too." This went on for 3 hours: nine sandwiches, one at a time. This wouldn't seem so bad had all of the crew members not been up for 23 straight hours, had to sleep in our uniforms, in our stations, just to be ready for any request at any time of the day. This was further complicated by the fact that the air conditioning was down for two days and the galley was in the crew mess. Sleeping in your clothes in 100°F conditions with constant noise and interruptions does not make for a good rest. Keeping your attitude in check under these situations is a challenge, but there is not a lot of money in saying no in our business or tarnishing someone's $400,000 vacation with tales of your pain and suffering. It is only for a few weeks, and if not, you can find a mutually convenient time to separate. At the end of the day and years down the road, the stories you have are worth it.

There is a flip side to yachts and celebrity. In summer 1999, the Backstreet Boys were the biggest act on the planet. I have never met or worked for them, but while circling the Mediterranean on a beautiful new 124-foot Delta motor yacht, owned by an old French Canadian man, we found a way to capitalize on this. We were a young, good-looking crew, and our owner was not a fun person. Every time we pulled into a new port, we would go to the dock master's office and in the background we would somehow mention the words *Backstreet Boys* louder than the other words. Within an hour and a 100-kilometer radius of a particular gorgeous Italian seaport, every girl who thought she had a chance with the boy band would show up in her sexiest outfit and hang on the marina fences—four feet deep. We would wear sunglasses and wave out the windows and blast their music on the outdoor speakers. At night, marina security would have to split the crowd at the back of the boat so we could hit the clubs; these were epic times! All the while, the old hardware store magnate slept, oblivious to our schemes. Sometimes a little fun is okay.

CHAPTER 19

The Dark Side

The race to utilize foodstuffs while they are still at their peak, late last-minute deliveries, and hungry diners waiting on their next course can easily max out the average person's threshold for chaos. Everything in your work environment is time sensitive. Add to that a long workday over a hot stove, and it does not make a recipe for long-term success. By its very nature, being a chef is stressful, but it is what you signed up for.

Chefs are artists. They sacrifice for years to learn a craft, and it takes some kind of education, either at a school or with someone investing their time and sharing their knowledge with you. During your rise to unsupervised competence, your contributions to a business are generally not considered worthy of a large payday. The business is taking the lion's share of risk and is benefiting from you as only a laborer. Therefore, most cooks work more for prestige and a title as they work their way from dishwasher to prep to working on the line and hopefully one day become a sous or executive chef. Cooks work crazy hours under pressure in a hot environment and then go home to an apartment that is usually full of people; when you are young and paid very little, roommates are a given. Bringing work-related stress home is nearly unavoidable. I can't tell you how many times I would lie in bed at night while the imaginary ticket board on the line in my head piled up with waitstaff barking out demands that their table be plated next. It takes a while to realize that you are in a dream state and that it is not real. I bet I am not the only one to whom this has happened. Most cooks cannot afford (either via time or money) the ability to escape to their own country house or peruse exotic hobbies to distract them from their job. With usually only one day off to sleep in, do laundry, and maybe catch a movie, most cooks turn to affordable, easily obtained methods of escape: booze, drugs, and cigarettes. Throw in a few tattoos and some piercings, and you have told the world that you are a unique individual. Sound familiar?

Once you become an executive chef, you will probably already be accustomed to the lifestyle that you have set for yourself. Now you will be able to afford better booze and better drugs and to stay loyal to your brand of smokes to show that you have not let success go to your head. At some point you will date someone not from the hospitality industry and get to see how normal people live, and this will make you want more. It is statistically rare to find cooks well into their forties, but the chefs who do make it are those who have paced themselves and have either accepted their fate or (if they are lucky) found a healthy lifestyle and truly love what they do.

When you go to work in private service, chances are you are not covered in tattoos, you probably do not stink like an ashtray, and you can probably operate sober. But that does not mean that the stress level goes down—quite the contrary. Travel is exciting, but as you prepare to board a plane for the 500th time in your career some of the magic is lost. Working for one of the world's most important or demanding people does not make for a relaxed environment. First of all, you almost never work alone. You may be very competent and in control of what you are doing in your department; however, the person serving or cleaning up may be a nervous wreck. You humming and gliding around the galley truly happy with your job and what you are doing makes them feel even worse about themselves or brings out how incompetent they feel, even if they are not. Often you may not have very much in common with the people on your team. Entirely different social-economic class and privilege barriers exist. Odds are that the language in which you communicate is not their or maybe even your first language. I remember working on a boat where all communication in the owner's areas was to be in French. I do not speak French fluently, but *bon jour, s 'il vous plaît,* and *merci* quickly became natural.

Not going home at night is a big deal for many. Security and convenience concerns make living-in a must with billionaires. For yachts, it is obviously a given unless you have a seasonal homeport, and the boat doesn't do a lot of overnight travel. When you are department head, you sometimes get ridiculously nice accommodations: one such situation that stands out to me was in Lyford Cay in the Bahamas. I was given a multimillion-dollar house on the golf course right next to Sean Connery's home. My house was nicer than his, and I had my own full-time maid. I would come home from the big house at night, crank up the Bang & Olufsen stereo, light up the pool, pour a cocktail, and go for a swim. I lived like a reclusive king. Breakfast awaited me every morning at my pool, and I came home to an immaculate home, bed made and clothes pressed. I had a new car issued and earned a healthy salary. Never saw this coming when I was in culinary school. I had a Boston Whaler with twin 150s on the back that I could cruise around the island to Atlantis and downtown Nassau on my day off. There was only one problem: I was background checked, drug tested, and fiduciary. Anyone I met along the way was not. I could not share this with anyone. No one was allowed through the gates into Lyford Cay. You had to check in at the gate; I was always joyfully greeted with a, "Good day, chef!" But if others were in the car, I would have to give names and IDs of all the people in the vehicle. I was not allowed to talk about the family or my situation around the island, had nothing in common with my coworkers, and enjoyed privileges far outside of their pay grade. At first I loved all the perks, because they are a great part of the job and definitely offer a distraction and shield from feeling sorry for yourself. However, in an area 40 minutes from downtown and securely locked away behind high walls and armed guards, a chef can get lonely. If I had a wife or long-term girlfriend who could pass all the background checks, too, I could have negotiated that she come with me. The demands of being the chef to a billionaire are that your attention is directed at taking care of the boss's needs. This is not an easy pill to swallow for a spouse or lover. Food service jobs are notoriously hard on personal relationships; in private service it is magnified. The chefs who do find a soul mate often do not last long-term in private service unless their partner is also a department head for the same employer.

I did eventually meet a nice lady at a cocktail party in Nassau and in, a desperate effort for companionship, snuck her into my Lyford Cay multimillion dollar accommodations. I could not tell her much about where we were going ahead of time, and she freaked out—it was too much for her. I made her a few strong cocktails, but she kept flicking cigarette butts everywhere. I had not even seen her smoke before but had to keep picking them up as they were evidence that I had had a visitor. She just kept asking a thousand questions that I was not allowed to answer. She could not relax, and eventually I had to take her back to town. All I wanted was to connect with someone; I needed a human touch and someone to swim with. I just wanted my house to feel like a home, even if just for one night. I have never felt lonelier than after dropping her off and knowing that not only was she not going to call me but also that she was going to talk about it to everyone on the island—and I was nice to her. Some situations are just too much for some people to take. From then on in that job, I had to settle for taking joy in my work and knowing that it was not a time to live life to the fullest—just in luxury. You see, after three days or three months, you can get used to anything. Mansion or shack: after a while it is just this is where I shower, this is where I sleep, and this is where I eat. The trick is not to get too used to the shack.

Loneliness, being part of an elite team, traveling with a high-profile figure, and traveling internationally all create stress, even when things are going well. How do you deal with stress without self-destructing or at least showing it? It is different for each of us. Smoking is generally a limiting habit in the private chef world. On an agent's questionnaire, answering yes to a question about whether you smoke is a kiss of death, but I know many chefs, particularly females, who have not getting caught down to a science. They do not smoke in confined spaces where the smoke permeates clothing and have learned all the blind spots on an estate or in the marina. A lot of staff on the yachts turn to working out, as many larger boats have gyms crew are often allowed to use at odd hours. Staff can afford a gym membership near a homeport or an estate, and it is not uncommon to see them jogging or running through the marinas and port cities. This is awesome but not always an option when at anchor, working on a small island, or during prolonged stretches of having guests in full service. Many practice internal escapism with an iPod and headphones or losing themselves in a movie on their laptop. This is healthy, too, but I remember it taking three days to finish a movie because I had only 30 minutes to get back into the plot here and there while being called back to service. I could not wear headphones as a department head, which serves as the communication hub at the boss's beckon and call. I have noticed a trend that has popped up in the social media era and now that Wi-Fi and smartphones have become so accessible. Whole yacht crews will sit around the crew mess table with laptops talking to friends all over the world and back home instead of to each other. This is a double-edged sword. It is wonderful to have good communication with your family and friends. It is great to share your unique journey in real time and not have to explain what you do when you go home for visits. Where this gets tricky is when you start to share the ship's information with others who should know nothing of your boss's itinerary or who they have onboard or over to the house as guests. Loose lips sink ships, and this can literally be true. You can get your boss or the crew killed. It is very rare that someone becomes one of the 2,000 richest humans in the universe without making a few enemies. I once worked for the man behind one of the largest security companies in the world. I will never forget him telling me, "Friends come and go, but enemies accumulate." You may be caught up in your own world and the politics of the asset you work on or in, but you will never know how complex the boss's world truly is. You do not have enough letters in your alphabet to describe the complexity of having 50,000 employees, thousands of tenants, or 500 people in your private staff. Best not to tell where you are going; instead, post once you have left.

These jobs are very exciting when they start. A new journey, arriving in a new destination, the first time you walk into a fabulous estate or board a floating masterpiece can really fuel the tank. This is the fantasy world that so few get to see, much less carry out in their lives. Over time, pressure and politics slowly build stress. It's like gaining three to five pounds a year; you don't notice it day to day, but after a decade it makes a big difference. In this case, you could be talking weeks or months. When you join a private, live-in service team you give up a certain amount of your freedom and privacy. On the yachts, we live in small, albeit nice, cabins. This is "personal" space (usually shared with another person) and is slightly smaller than a jail cell. Humans were not meant to live like this for long periods of time. This is probably why there is so much turnover in yachting. We work long hours, and carrying a sleep debt is a way of life. Eventually, it is human nature to try to eke out something of a normal life, even under these conditions. Crew hook up, and people establish friendships, which is all good and healthy until something inevitably goes wrong. When a romance goes south, it can be devastating, particularly when your partner chooses another lover on the same boat. I have personally worked on two yachts where crew committed suicide on the boat. Thankfully in both situations it was a few years after my time on the vessel. In a worst-case but real scenario played out in the Caribbean, a female crew member lost it aboard a sizable yacht and was walking around the aft deck of the boat with a gun to her head and threatening to end her life before local authorities ambushed and disarmed her.

But violence is not always the outcome. Some chefs turn to pornography as a stress release, although lack of privacy makes this difficult because of close living quarters. Many crew have caught someone or been caught pleasuring themselves; while it's embarrassing, it is a natural impulse, and the crew will all have a few laughs about it for a day or two. One circumstance in this regard was when after a nice evening out a couple came home late at night and passed through the country kitchen on their boat to get to the crew mess. They told me that they found the new chef masturbating on the settee to pornography on the TV. I said, "So what: you surprised him." They said that wasn't the problem: he didn't stop and instead just stared at them.

I remember when I was young and living in the Virgin Islands, my friend had a horse trail ride business and offered a summer pony camp for kids. One day I saw the thoroughbred I rode with his top teeth on the fence stretching back and gulping air; she said he was cribbing, or gulping oxygen to give himself euphoria. She said that he was showing signs of stress: with all the little hands and people on and around him all day he was out of his comfort zone, but it was also what paid for his care. In private service I have seen many people crib in their own way. Every situation is a microcosm that can bring out the best or the worst in you and can result in behavior you would not normally do in other situations. There are many instances of improper interactions with guests. On one occasion an older male chef imposed himself onto a private jet being used by a famous race car driver's wife and her girl-friends and proceeded to creep them out for their entire flight home. Many a crew have crossed the line by hitting on the boss's wife, girl-friend, or daughter. Making an owner or guest uncomfortable is a high crime in our industry. Because of random testing and tight quarters, IV drug use is rare. However, club, recreational and prescription drugs do make their way through the ranks, causing friction and often put-ting a program at risk. Scoring drugs to gain favor with the boss or guests is not unheard of. Once, a male chef upon scoring cocaine for a lesbian couple chartering a yacht was invited to their quarters to join the party—this they didn't mind. Later, emerging from their in-suite bathroom nude and pleasuring himself did not go over so well. He now works in another field.

Thankfully, I was not the perpetrator or witness to any of this, but it does not mean I am immune to acting out in stressful situations. I was once the chef and estate manager for an estate in the Georgica area of the Hamptons and was given an incredible housing, salary, and a brand-new $94,000 supercharged Range Rover as my staff car. I had hired a pretty girl from the yachting business to work with me, and we had just hired a nice couple from Ron Perlman's estate next door to carry out indoor and outdoor maintenance. The boss was a billionaire and well-known superyacht owner having built or refit boats totaling over $100 million. They were known as great people to work for. While building a home in the neighborhood they were renting a house with ties to the Beatles that had just been completely renovated. I was the purser and paymaster; the family was never in residence during my whole tenure. They told me that they had no credit card for the house yet and asked if the staff could just put daily purchases for the house on our credit cards and they would send a courier with cash to reimburse us. The boss basically said, "We rented a house in the Hamptons; can you all drive up to New York and set it up for us and we will pay you back?" We had accounts at many places but not for everything we needed. The staff maxed out their credit cards in good faith, but the money did not arrive as expected. Naturally, they became stressed out over not being able to take care of their own personal obligations back home. The boss was not sending enough money to pay everyone, and I could pay them only partial salaries. They would drop in saying that they would be bringing cash and as they drove down the driveway, and I would run out with a bag of receipts and a ledger.

They would give me a few hundred dollars out of their pocket and send a courier the next day with only half of the expected money. This became very stressful for the gardening couple as they had tenants of their own moving out and new ones moving in with only three days to replace the floor. Now that they maxed out their Visa card with $5,000 on the boss's house, they had no way to pay for their floor. By this time, the girl I had hired was asking me what I had gotten her into, as now she too was late on her bills. The lesson here is never loan rich people money. You will get paid back, but it will be on their schedule, not yours. We had food, gas, and a nice house for sure, but that was a veneer to a mounting problem. I don't think the boss was evil, just careless. I started having to be a strong man in conflicts between them and the homeowner, the rental agent, and the townspeople.

One day after picking up the four-wheel drive beach passes for the fleet of family vehicles, I kind of lost it. I drove out to Maidstone Beach and unleashed the fury that Range Rover had to offer. I wrote my name in the sand with rooster tails of sand coming out from every tire. I had the radio blasting; the sun was setting, and the tide coming in strong. Nothing like burning fossil fuel in a truly powerful and expensive machine to let it all out—I drove through the surf as fast as I could, with the pinks and purples of the sky reflecting on the wet sand. I skidded to a stop to take it all in: wrong thing to do. When I tried to start moving again, all four tires buried into the sand until the chassis was supporting the whole car resting on the sand. Now, I am a billionaire's problem solver, and my mind works fast and without limitation; I was screwed. The Hamptons at the time were a mixed bag of the best and latest in technology and simple country living. Cell phone reception in certain areas was nonexistent and depended largely on the cellular carrier. There was no signal, and the waves were lapping against the doors at this point. I had AAA, but even if I could have called them, the car was hundreds of feet down the beach on soft sand. And in that area, which was a ghost town midweek, you were lucky to get an old-fashioned tow truck the same day. You are supposed to deflate your big, fat, four-wheel drive tires to allow for operating in the deep, soft sand. I could not do this with the oversized rims and low-profile tires that came with the vehicle. I lasted as long as I did only through planing over the sand at blistering speed. I made the hardest decision ever in an effort not to end my job and stunt my career: I planned on walking for miles back into the center of East Hampton, catching a movie, and then reporting it stolen when I came out of the theater. (I have never shared this before.) I made it as far as the pavement when out of nowhere a bunch of high school kids came charging up the beach in an old Toyota pickup truck with the confederate flag painted on it. They said, "Holy shit! Is that yours?" I lied and said yes and that I was going to look for help. They handled the whole thing: found a chain, manned my vehicle, and saved me from a nearly six-figure loss of one of the boss's assets in a selfish, careless act. Not something easy to recover from. I let them beat the hell out of the Rover for a while and then thanked them and gave them the $200 cash I had on me. After hitting the full-service car wash, I ate my supper alone and went to bed early. Oddly enough, I did feel better afterward.

Many of us would cope with the stress by having crazy out-of-the-box experiences off the boat or out of the area of the estate that we were working on. In some countries and islands whole crews, male and female, will go to a strip club until well after the sun has come up just to get in an environment with less rules and less judgment (nothing seems to count there, and we don't bring it up later). For the boys, visiting brothels is considered completely acceptable. When you live in, you cannot bring strangers home, and you rarely have the time or are in a place long enough to properly court a local girl. If you do, you have nowhere to take her as you will have to report to duty at some point.

The ladies of yachting are no innocents either. I remember pulling into Annapolis on a day where the whole Naval Academy had a special night out in town where everyone had to be in a uniform. A sweet, petite, adorable, bible-thumping South African stewardess who rarely ever left the boat admitted after a few drinks that she grabbed a random sailor off the street and had him take her behind a dumpster. We were all jealous of that sailor. Another occasion that comes to mind is when we had this slender, pretty Australian former cruise ship dancer as our chief stewardess. We worked for a Swiss-French lawyer/TV presenter who was loud, passionate, and rarely happy. He fired people left and right, and that stressed her out. She hired on a young, hippy type who spent all her free time traveling to music events like Coachella and Burning Man. These two had nothing in common, and the chief stewardess spent most of her time correcting her in frustration. When the trip was over, the whole crew went out drinking in Grand Case on the French side of St. Martin until 4:00 or 5:00 a.m. On the ride home in a taxi van, the two of them sat all the way in the back, and I was ahead with a few other crew members in front of me. The radio was blaring, and somewhere over the hills on the way back to the marina I looked back and saw the two of them slowly, deeply, and passionately making out. It was blisteringly sexy, and I never told anyone about it; both had their eyes closed, so they never saw me peek. Both had long-term boyfriends, and the hippy flew out the next morning. I doubt they ever spoke again—just a brief escape from reality.

Private service is a sea of temptation and deprivation all at the same time. Inevitably, where there is money, attractive people will show up—either as immaculately groomed, couture-draped guests who seem to smell better than regular people or as international, fit, young staff. I have shared cabins and bunk beds with some of the most decadently gorgeous young ladies with the sweetest accents who would climb over me to get to their bunk wearing only satin underwear. It is almost physically painful after going a few months without sex and having to try to fall asleep only 18 inches away while breathing in their fertile pheromones. A few hours in a brothel with young ladies fighting over you and your American dollars goes a long way and is an ultimate escape from the stresses back at work (I am told). Like the spit valve on a trumpet, empty it every so often and it gets easier to play.

There came a turning point in private service in mid-2000 when crew stopped hanging out only at the crew bar at the end of the dock and doing local side trips on their days off. As newly built yachts grew in size at an alarming rate, so did salaries and stress levels as numbers of crew and expectations grew to meet the demand. I remember renting Harley-Davidson motorcycles and taking local day trips around the area and being satisfied. Suddenly, crew was flying out of St. Martin to ski in Colorado if ever guaranteed a long weekend. We started not wanting to see any other crew in our off time. This was sad as then you associated your fellow crew only with work and not seeing the fun side of them. Crew and staff tend to put a lot of pressure on each other and themselves to perform at a high level in an effort to secure higher tip levels and favor with the boss. I can understand not letting your guard down because people have lost jobs over their behavior outside of work when it was reported back by their own coworkers. It is nice, though, when the whole crew goes out to a happy hour or a nice dinner and just relaxes without any drama.

Some situations are just beyond your control. I once got called in for a simple, three-day trip on a smaller yacht for a wealthy TV funny-man and his family who wanted to unwind after he completed a movie project. Nice crew, nice family, another couple, their child, and three nannies for a total of four children. Less than an hour into a cruise around Shelter Island, the youngest child (under 2) got his tiny finger caught in a large flat-screen TV on a hydraulic lift, which his brother was operating while under the supervision of all three nannies. The poor little fellow was hurt. The captain was from another part of the world with a lot of protocol and not used to celebrities with their sense of immediacy and "break the rules" in case of an emergency mentality. The captain took an extra 45 minutes to do a proper mooring while the boy was in pain rather than dropping some fenders and going side to side with a docked yacht to do a live handoff to get him into a waiting ambulance or dropping the tender to run him and his mom into shore. Mom was livid, and I didn't blame her. The funnyman was to join the party after the cruise. The other guests did not know what to do and eventually left, as things were so chaotic at that point. This was a high-profile charter. Lots of industry eyes were on us, and though it was not our fault about the accident itself Mrs. felt that the captain did not care about their safety or comfort then she no longer wanted anything to do with this vessel or trip. This was a $50,000 weekend, and there were no refunds. We all sat around the crew mess wondering what was going to happen. Phone calls went back and forth between the charter brokers on the West Coast to the yacht's agent in Florida all night long. The captain would give us updates as they came in. We started out as a warm fun group put together to give easy joy to an A-list celeb, and now we were speaking awkwardly to not at all. Theoretically, they could return at any moment and we were at anchor, so no one could go ashore. The crew mess and cabins were not designed for spending a lot of time. You could cut the tension with a knife, and it brought out the worst side of others and me. There was no escape—no TV or computers, we could not drink as we were technically on charter, and we were all cramped in tight together as the rest of the boat was highly detailed.

It is one of the worst feelings when the guests no longer want to be there or in your presence. In my 100 or so trips, it has only happened a few times, but each one stays with you. It is rarely started by something the crew did, but more like the boat is not as advertised. The captain may not respond to unforeseen problems well, or the charter guest may breach the contract (e.g., drugs found onboard). Whatever the case, fingers get pointed, there are big money and reputations at stake, and the feeling takes a while to flush from your system.

Finding your own path to deal with this level of stress is a personal journey. I can only say this: when you have gained a significant amount of weight, have to go on anti-anxiety medication as a way of life, or find yourself not acting like yourself, get out. Living for the wishes of one person—to whom you are not related—is not a normal way of life. It can be awesome and decadent for a period, but it would be a disservice not to mention that the stresses are there. When it stops giving and starts taking from you, it is over. How you walk away is easier said than done.

FIGURE 19.1 THE YACHTS AND MANSIONS ARE THE CANVAS, BUT THE PEOPLE ARE THE PAINT. MY GALLEY TEAM AND ME DOING THE THRILLER DANCE AFTER A SUCCESSFUL $1 MILLION CHARTER; THEY ARE NOT ALWAYS SUCCESSFUL.

CHAPTER 20

Your Career

When you are starting a career in the culinary arts, you will look for a break to get into the game. Someone somewhere has to take a chance on an upstart with little to no experience. Once in, you will most likely have a string of jobs that are just to get some experience, find your strengths and weaknesses, and learn how the whole power structure of the industry works. Once you have built enough of a track record, you will become an established commodity such as a line cook or sous chef, and you will now have more leverage when seeking out jobs. A hotel or restaurant job will usually have some room for upward mobility, and you may spend years in one location climbing the ladder. Someone once told me that it is important to be where you are supposed to be in your career and that if you are not fitting in, happy enough, or passed over for promotion you will need to be brave and make changes. It is important for you, as a chef, to keep growing. Growth in knowledge, experience, and power are crucial throughout your career. Being stagnant can kill a creative person. I believe that to become a great chef you will need a varied collection of experiences to draw from and to base decisions on. You do not need good or bad information—you need *all* information. It is important to work in a failing business or two on your way up. It is also important for you to work for a bad chef or two. This will give you the perspective to see when things are going south and to either get out early or help make changes. Working for great restaurants will show you how things can be if enough smart people are making good decisions. I know a group of petty criminal/con artists from a bar I used to visit in Florida, fun likeable guys who would buy the beers just to have an audience to hear their stories of how they got the better of people with their well-thought-out scams. I learned more about how to avoid getting ripped off from them than I did listening to most successful people talk about how to be successful.

A string of jobs does not make for a career. A real chef is a chef whether they are employed or not. The term *chef* is French for chief; it is a person who is in charge of all people and assets within their domain. Running a kitchen does not make you a chef; a diploma from a fancy culinary school does not make you a chef; taking a test from an industry association does not make you a chef. Experience and a body of work are what make you a chef. At some point in the game, if you have taken on enough responsibility for long enough and under that title, you will be a chef. You may be a good chef or a bad chef, but it will be part of the fabric of your life and your identity. You will know you are a chef when you are considered a chef in the eyes of your peers. To be a real chef you should know all forms of rudimentary cooking and baking; know how to hire, fire, and team-build; have excellent communication skills; know how to serve; maintain all the equipment in your domain; know how to purchase and steward; understand a profit and loss statement; keep the employees safe; and, above all, be able to connect with the people paying for it all.

Okay, so you are a real chef; now what? A career is a long journey with ebbs and flows. Rarely does someone sail directly to the top in solid trajectory. If it happens to you and you are awesome at it, great! But most of us start at a safe level where entry-level mistakes will be left behind and won't follow us to the top. I have helped many good people—both chefs and stewards—get into the business. I tell them that they do not have to take the first job that comes along and that there is a learning curve to private service. Sometimes they listen, and sometimes they do not. One young lady comes to mind: a front desk clerk I met at a hotel where I was staying in Santa Barbara, California. She was quite presentable, but more importantly she had the best customer service skills I had seen in years. She could read people and meet their needs in a very efficient manner. She could carry out her tasks professionally without seeming like a corporate robot. I offered to help get her on the yachts, and a few months later she flew into Fort Lauderdale. I already had her register with all the agents, and her paperwork was in good order—too good, in fact. The agents went nuts over her, and within hours (not days), one talked her into a large, high-profile boat with a very serious program. I said, "No, don't take it. You are 20 and should have some fun with a nice crew on a smaller boat first." She felt she had to take it and did. Just 17 hours after landing in Fort Lauderdale, she was on another plane to meet the yacht. She was one of the best workers they had ever had, they worked her hard, and she became the personal trainer to several crew. The owners liked her as well; who wouldn't? They wanted to start to train her to be the chief steward on a $40 million yacht so that she could take over when the current chief was leaving in less than a year. She was performing very well, but the stress of all this before her 21st birthday was too much. Two months in, she quit the business altogether and headed back home. A year later she took a summer job on a smaller yacht as chief steward and hired a friend from home to be her second steward; they had the time of their lives. A career should start at a comfortable level, and once you get your head around the situation you can begin to ascend.

Okay, you are a now a private chef and you have a modest estate or manageable yacht or two under your belt; now what? You are going to have to ask yourself, What am I doing this for, and what do I want to get out of this experience? Do you want to make a lot of money? Do you want to get laid? Do you want to become well known or respected in the private chef world? Do you want to travel? Do you want to push the limits of your culinary potential?

It may be one or a blend of these motivations that lead you down this path, but the sooner you figure it out, the better. There are different focuses and ways of going about achieving any of these goals. A career is best based on goals. I like to think of plans as short-term and goals as long-term. You see, the trouble with a plan is that you either get let down or you get exactly what you expect with no chance for better. When you have a goal, you make active decisions along the way based on that goal. You take the opportunities that help you achieve your

FIGURE 20.1   CHEF NEAL IN VENICE, ITALY, 2001.

goal and do not waste your time on ones that do not. You have to make these decisions on your own. I remember a time when there were more chefs than chef jobs on the boats and a crew agent would say, "You're not ready for that size yet." Just a few months later when there were more jobs than candidates, the same agent would be saying, "C'mon, Neal, you have got to step up sometime; you can do this!" Take stock of your peers. After a while you will be able to assess whether or not you are as competent as or more than the guys at the next level above you. You do not have to stay in a lesser position out of respect for your craft. Remember that when you are in a pack running from a tiger, you do not have to outrun the pack—only the tiger.

It has been my experience that when someone works hard and stays focused on their goal, other people take notice. Someone may admire your commitment and discipline. You may even be offered something more ambitious than your current goal—something that you may have not even known was possible. Do not be afraid to switch goals as you learn more about yourself and the industry. We don't always tell the newbies about the really good stuff.

Knowing who you are and being realistic about what you have to offer is important. You need a tremendous amount of ego and self-confidence

FIGURE 20.2   CHEF NEAL, HUGE RED SOX FAN, IN THE BOSS'S SEATS AT FENWAY PARK.

to take on an executive chef position in a major program. There are times in a long career that other issues may require your attention: a sick family member, a health issue, a serious romance, or just plain burnout. It is okay to take a step back and run a low profile for a time. You reap what you sow in a career, but if you take a season off or take on less responsibility for a short while, you can come back when your batteries are recharged.

Estate positions tend to be longer term and are mostly done through specific agencies or local word of mouth almost exclusively. You are not in the public eye that much, and barring a seminar or a random meeting in a grocery store, you will not be interacting with that many other estate chefs in your line of work. In the yachting business, you are on stage full time. Every marina you are at will be filled with your peers. Parties, networking events, and boat shows go on all throughout the year. When you live in a place like Fort Lauderdale, it seems like every other bartender has a captain's license and knows all your peers. You never know who you are sitting beside. Never talk about available jobs in public in any yachting capital port; the grandma at the table next to you probably has a friend, neighbor, or grandchild in the business. If someone overhears that there is an open gig, word will get out fast and people will be bombarding the vessel with CVs (resumes). The captain does not have to pay a commission to an agent if they find a good candidate on their own.

I am convinced that yachting is the third most image-driven business in the world after modeling and acting. You do not have to be gorgeous, but you need to look the part. When it comes to interior staff, you have your workhorses and your show horses. You get more money for a show horse. It pays to be fit, it pays to be well groomed, and it pays to show that you have been successful to date. When going to any yachting event, barring a beach barbecue, dress to impress. Clean, pressed clothes and good shoes are a must. They can be boat shoes, but not the ones you wear to clean the hull. It is human nature that people would want to hire people who take care of themselves and show respect to others by making an effort to present themselves in a dignified way. If you are at the top of your game, you will be hired on your track record and do not have to try as hard; however, on the way up, I used to like to be the best-dressed guy at the party. It also helps if you arrive with quality people; tell me who you live with and I will tell you who you are. Even if you are showing up with people who are not in the business, bring people who have something of interest to the event: good-looking, fun, polite types attract attention, and if they are good people they will say that they came with a great chef and will point you out. While these types of friends/guests are mixing and mingling, they are marketing for you. The more people talk about you at the party, the more your message gets out: you are a chef, and you are available. When you attend these industry events, it is a good idea to meet as many people as you can. Try not to get tied down to any one group or individual. Yachting can be lonely and stressful, and many attend these events to make friends and feel a part of the scene. Be nice and polite, and enjoy yourself. In our industry, we live with our coworkers, so we judge them not only on how they work but also on how they play. When you meet people you like, arrange an after-party somewhere—"We are all meeting up at Waxy O'Connor's after this."

The party itself is for networking, so you can take more time to get to know each other at that point. Allies are everything in an unregulated business. You need people to have your back and spread the good word about you. You never know where the next job is coming from, and since there is no sure-fire way to get that dream gig you need to play the odds.

When it comes to getting work in the private service industry, you need to stack the deck in your favor. You can never be with too many agencies. Agents are not the only way to get work; however, it is the only part of the game that you have any real control over. It is like playing the lottery: you cannot win if you do not buy a ticket, and the more tickets you buy the better your odds are. When I was active there were as many as 20 agencies worldwide with which I was registered and checked in with regularly. An agent is a person or business that markets itself as being able to supply qualified candidates for a fee. The problem here is that anyone can call themselves an agent; there are no regulations. In the 1990s, an agency was most likely a few former crew who were well known and retired from active life at sea. They knew the captains and the crew, and they played matchmaker. They ran their operations from a rented storefront with cheap furnishings, a phone, and a fax machine. I liked this period, because you walked in and talked to the boss without pretense. The whole value in the operation was the agents themselves—they actually knew the players. They were very hands-on back then. If you needed a haircut or had an attitude problem, they would yell it at you straight to your face, and you fixed the problem before it got out of control.

When the business got larger and more organized (but still unregulated), brokerages and management companies realized that there was money to be made in crewing up these new larger boats. They bought many of the agencies and moved them to fancy offices in low-rise buildings; now the crew in their pressed polos and khaki shorts were riding the elevators with the folks in business attire. Soon spouses of department heads and anyone who thought that they knew enough people in the business opened a crew agency or marketed themselves as an independent agent. This really made it confusing and frustrating for crew. You want agents to be interested in you. It was tough, though, to discern who was the real deal and who was wasting your time, whose spouse was helping them look successful even though they actually rarely actually placed any crew. More established agencies now had you set an appointment a week in advance just to meet for 5 minutes face to face. The other frustrating part was that now everything was being done online. Crew now had to deal with poorly designed and misfunctioning websites that wasted your valuable time between jobs. You were never sure that your check-in was actually seen by anyone.

There is a lot of nepotism in private service; people tend to surround themselves with people that they like rather than the best candidates. Several times agencies hired pretty receptionists right out of college and straight off Craigslist. It was often their first job; they had never worked on or even spent any time on yachts and did not seem to understand who we were or how urgent some matters were. They guided us to a bank of computers on the wall and told us that we had to handle it online. What became worse was that that the receptionist would end up being promoted to a crew agent and was now in charge of deciding which crew got put up for big jobs. Things have gotten better, but the moral of the story is to not put all of your eggs in one basket: register with all agencies. Some of my biggest gigs came from the smallest agencies or from agencies I had not heard from in a while. When it comes to the agencies for yachts, the squeaky wheel gets the grease. I used to make the rounds of the local ones twice a week. You will never do better than a face-to-face meeting, and sometimes that is in an elevator. The biggest single advance in my career came when I ran into an agent in the parking lot. They said that there was not much going on except for this one really intense program heading to Europe in a few days, but the captain was known as a strict micromanager. It was a flagship of the best yacht maker, and that's all I cared about. Micromanage the hell out of me! I wanted to be on one of the nicest yachts on Earth.

In between gigs, I spent an average of an hour a day checking in with different agencies online or in person, going to industry events, or having lunch at a crew hotspot. Make sure that you do take some time between jobs to enjoy your life because you won't have as much freedom or privacy once you are hired.

In yachting, you kind of hover in the industry standard range for salary until you are at a point where you are ready to make your move to take

FIGURE 20.3   THE BEST PART OF THE JOB: THERE ARE ALWAYS BEAUTIFUL WOMEN
AROUND TO HIT THE TOWN WITH.

FIGURE 20.4   CHEF NEAL MOTORCYCLING FRANCE AND STUDYING WINE IN 2001.

on more responsibility. This is when you have successfully become a member of the pack of recognized yacht chefs and have established a responsible track record. Now is the time to separate yourself from the pack, but it is important to do this when you are actually ready, not when you feel ready—and to know the difference. Big boat chefs are treated like rock stars in our industry; billionaires praise you, magazines write about you, everybody knows your name and we earn serious cash. We have jobs of creativity in an industry of regimented discipline. A lot of people aspire to become one of the brand-name chefs, but if you do not have the right stuff it is never going to happen. Some chefs have the right stuff but do not believe that they are ready, even though they are and the clock is ticking. Some people date for years but do not feel ready to get married; at some point, you just commit to it, plan a wedding date, and become a married person. After a short while it becomes your new identity or the next chapter in your life. If you are ambitious, you must not stay in the average range too long. Yachting has a shelf life. It's a young person's game that consumes your whole life; it is a lifestyle career that requires stamina and focus. It is not the few times you do extremely heavy lifting that will break you; it is the lifting at all times of the day and night. Being a yacht chef is very physically and mentally draining. If you have a wife who is not going to do this with you, you have children, you have any physical limitations, if you value freedom and privacy over global adventure, or you are just not fun to be around, do not attempt this as a career. There are only a few thousand of these elite jobs operating anywhere in the world at any one time. If you do not have your head and body fully in the game, the odds of you getting one of these amazing gigs is not in your favor. If you have become a successful, well-functioning yacht chef and have decided that you are ready to go for one of the bigger jobs, realize that the larger the yacht, the fewer of them that there are. These jobs pay more and have stratospheric perks. These types of jobs do not turn over often, and there is no room for error at this level. I can tell you that working freelance on many boats will make you a better chef—a much better chef—however, at the highest level superyacht and gigayacht captains tend to be highly qualified, no-nonsense types who are looking for impeccable references and longevity above all. Smaller boat captains value their reputations, and even if they give you a nice reference letter

FIGURE 20.5   CHEF NEAL COOKING, COPPER POT.

when they talk captain to captain nothing will be held back. I can say only that when you have enough experience and your captains and peers believe in you, go for it. It feels good when you reach the next level and have success.

There is a point where you will be financially stable enough, will have met the right someone, or will want to start a family, and you may consider getting out before the industry stops giving to you and starts taking. The problem with yachting is that is a very hard career to retire from. Years ago, I flew into Fort Lauderdale to participate in a large yacht cooking competition. En route to the venue, the taxi driver and

FIGURE 20.6   NEAL ENJOYING A DAY OFF IN AUSTRIA, 2001.

I struck up a conversation about the event and the yachting industry. The driver made the most poignant remark: "Wow, what do you do after all that—just fade away?" Elegant and frightening at the same time. Most people think you are telling a deluded fairy tale when you try to explain the life you have been leading. How do you move to the next chapter in your life without throwing away all of those amazing experiences and global knowledge? When any of us figure that out, we will write a book on that.

CHAPTER 21

How to Become
a Private Chef

So you have read this far. If you think you fit the bill and want to spend a portion of your career in private service, what's next? We will assume that you are a competent and experienced chef. Your first step would be to simplify and organize your life so that you are ready to go; things can move quickly in the private service world. The next step is to decide whether you are interested in estate work or yacht work because these are the two main venues for your career choice. If you are a stable and balanced person who likes a routine and the ability to have somewhat of a home life, you may want to try the estates. If you are an adrenalin junkie filled with wanderlust, don't mind cooking in an environment that at times ranges from miniature golf to a circus, do not mind living with others in tight quarters, and can handle plans changing minute by minute, maybe it's yachts for you. In short: ADHD = yachts, no ADHD = estates.

Good: now that you have made your choice, we need to get your paperwork in order. First off, do you have a *valid driver's license* and a *passport*? If not, getting both would be your first step. You cannot be hired to work for a billionaire if you cannot travel and someone has to drive you everywhere. There are exceptions, but you will be start-ing out at a huge disadvantage and the agents will not consider you a strong enough candidate. Agents put their reputation on the line with every job order that comes in. People come from all over the world to compete for these gigs; there is no shortage of bright, ambitious people actively seeking these jobs.

Next, if you are not an American, make sure that you have the proper *visa* to seek work in the country where you are looking. You will have to check with your local government on this one; you do not want to start your career out illegally. Imagine how upset the boss would be if they pulled their yacht into a new port and their new chef is deported.

The next item will be to get any required safety training. For estates, people like to see at least a current first aid certificate, as you will be on site in some large, intricate, and secure locations. If you are going the yacht path, it is time to sign up for your *STCW-95,* the five-day safety course covering fire fighting, first aid, sea survival, and social responsi-bility at sea. You will need this because you will not only be the chef but also an active part of a sea-going crew. You can look up locations and prices online, but for an STCW-95 plan on spending around $1,000 plus room and board. This is a great experience: you will train with a real fire department, put out real fires, and learn search-and-rescue tactics. You will also learn sea survival: actually be in the water, wear all the gear, and climb in and out of life rafts. You will learn how long a person can actually survive and how to play the odds in your favor to being successfully rescued. After this, you will learn how to handle politics aboard a ship and how to avoid and resolve conflict at sea. The final block will be medic-level first aid. When you are at sea, air evacu-ation may be hours away. If you need to be rescued by a commercial vessel far out in the ocean, it may even take days. You will learn a more aggressive level of first aid than the average land-based person will have to know. The course is good in that in an emergency you become a trained rescue asset rather than an unsure liability. It makes rescu-ers out of rescuees. In an emergency, you never know who is going to step up and make a difference. The STCW-95 also separates the serious would-be crew from the backpackers, who are looking only for short-term adventure and are likely to do this kind of work poorly and on the cheap. You see, you can actually work on a private yacht without your

STCW-95, but you cannot do any charters without it. You would have to be replaced by a qualified freelancer for charters. Not only is that expensive and a pain in the ass for the boss, but also charters are where the money is. A week's charter tip can easily be more than your week's wages, and you get both when you charter. It is like popping a nitrous bottle on your bank account. No good program hires a chef without an STCW-95; and the agents won't even interview you without one.

For yachts, you will need an *ENG-1*. This is simply a marine medical exam. It is a specific exam, so your normal physical will not do. Marine capitals like Fort Lauderdale have approved walk-in clinics set up for this, and it is not too expensive. Check online for pricing and locations. Some boats require a drug test, but they will normally arrange and pay for this on their end to limit liability and avoid forged documents. Do not underestimate how badly some people want these jobs. The 17th Street causeway in Fort Lauderdale is both the golden road and the street of broken dreams. I have done a lot of living on that street; I have received a Willy Wonka–level golden ticket so many times there but have also broken down crying as I climbed into a taxi with my tail between my legs.

As a chef you will also need a *sample menu.* This is not something you should study up on and create as a show of competence. This document is to show the boss, captain, or estate manager the style of food in which you are most comfortable cooking and to show your range within that comfort zone. A simple rule-of-thumb is to not put anything on this menu that you have not made at least 10 times successfully. The menu should look like a restaurant menu, as that is the format that most people are comfortable when reading about culinary offerings. It should be for seven days beginning with breakfast. One thing to include would be a continental breakfast of fresh pastries, yogurts, cereals, and a fruit platter. Then you can write out a different hot breakfast special item for each day. This is how it really is in most cases; therefore, it shows that you are at least familiar with the industry standard. After this you will do Monday through Sunday lunch offerings, such as filet of lemon sole with a blood orange beurre blanc, basmati rice and steamed asparagus, followed by a passion fruit sorbet. Do whatever you feel comfortable with, but an entrée or salad and a light dessert will do. For dinners you will offer three courses: a starter, entrée, and dessert. How about

Chef (Your Name) Sample Menu

Breakfast

Breakfasts will begin with a continental offering of:
Freshly baked Pastries and Croissant, Greek Yogurts, Hot & Cold Cereals, Freshly squeezed Juices and a Bouquet of cut Fresh Fruits
Hot breakfasts are available a la carte: no request refused.
Daily specialty offerings will include:
Roasted Leek and Chevre Frittata, Honey-Nut, All-Fruit Pancakes with hot Vermont Maple Syrup, Eggs Sardou, Florentine Soufflé, Foraged Mushroom Crepes, Homemade golden San Francisco Sourdough French Toast with a Raspberry Compound Butter, Lobster/ Fois Gras Omelets

Lunch

Monday: Green Salad, Tomato/Saffron & Local Fish Bisque with Crostini, Almond Lace Cookie Cup with Fresh Berries
Tuesday: Chilled Jumbo Lobster Claws over a French Nest Salad with Lemon Crème Dressing, Passion Fruit Sorbet
Wednesday: Crispy Paillard of Poulet du Bresse, Grilled Aubergine, Heirloom Tomato & Burrata Stacks, Lemon Parfait
Thursday: Beet Carpaccio with Petite Tournedos of Bison and Mangosteen, Cheese Platter
Friday: Japanese Vegetable Stir-fry, Pan-seared Patagonian Toothfish, Green Tea Gelato
Saturday: Kobe Beef Burgers with Hand-cut Steak Fries, Hot Fudge Sundaes
Sunday: Seafood Risotto, Toasted Panetone Wedge with Madagascar Vanilla Bean Gelato

Dinner

Monday: Vegetable Terrine followed by a Pan-seared Veal Chop topped with a Gorgonzola-Tomato Salsa, served with Cherry Pepper Spinach and an Olive-Basil Mashed Potato, Fresh Strawberry Shortcake
Tuesday: Seared Fois Gras with a Port Wine Blueberry Gastrique followed by a Grilled Loin of New Zealand Venison, Sauce Choron, Haricot Verde and Truffled Duchess Potatoes, Valrhona Chocolate flourless Torte with a Crèmed Rum Sabayon
Wednesday: Gruyere Soufflé followed by a Colorado Rack of Lamb, Sherry-Shallot Crèmed Demiglace, served with Swiss Potato and a Ratatouille, Black Forest Torte
Thursday: New Orleans Andouille, Chicken and Tasso Gumbo followed by Roasted Lobster Savannah served with Rizzi-Bizzi Rice, Commander's Palace Bread Pudding with Bourbon-Caramel Sauce
Friday: Chicken Consommé with a Quenelle followed by Chateaubriand with sauces Au Poivre and Béarnaise, Gratin Potato and a Bouquetière of Vegetables, Pear Belle Helene
Saturday: Moules in a Tomato Court Bouillon followed by Pan-seared Norwegian Fillet of Salmon, Sauce Citrus-Crème Fraîche, Basmati Rice and Grilled Asparagus, Profiteroles
Sunday: Caesar Salad followed by Lobel's Prime, Dry-aged NY Sirloin Steak, Sauce Roquefort Demiglace, Baked Potato and Broccoli Spears, NY Cheesecake with Bing Cherries

*Allergy and special dietary and requests honored

FIGURE 21.1   SAMPLE MENU.

a tomato consommé, followed by a pan-seared Colorado rack of lamb served with a sherry-shallot creamed demi-glace, Duchess potatoes, a bouquetière of spring vegetables, and poire belle Hélène for dessert. You get the picture; do one of these for each day of the week. You may also do breakfast lunch and dinner after each day rather than putting all the lunches in one section and the dinners in another, your choice. No one is ever going to ask you to cook that exact menu for the week or even anything from it; however, it tells the reader what kind of chef you are so they can anticipate if you and the boss will be a good match. If the boss is not a big meat eater and your menu reeks of carnivorous delights, it is best not to waste anyone's time or money. You can alter your style for a charter here and there, but if you are used to cooking standard fare and you sign up for a season with a vegan boss or crew, you will go nuts over time and you don't want that. Make the menu look nice, but do not get too crazy with the font. Many in private service use English as a second language, and cursive scrolls may work against you if they cannot read it. Be sure that you have your name on the menu and in the document title. We used to be asked for a separate document of photos of our work, but now that most of us have the capability to easily insert photos onto documents you can put the photos right on your menu page like a filmstrip. When it comes to your paperwork, looks count. The boss, captain or estate manager will most often not have the opportunity for you to do a trial dinner. They need to look at the menu to see if the choices are in sync with their needs, and the pictures will tell them if your food looks like it is up to par with their program. I cannot express how important it is to have good-quality photos in all of your documents. Keep a good camera in your kitchen even if it is a point and shoot, and when you see a plate that you are proud of take a picture. You want photos of your real food. Anyone can steal good food pictures off the Internet, but when you cannot reproduce the dish in person you will be deemed a fraud.

All of your documents should be Microsoft Word documents or PDFs. Word documents are the standard of our industry; however, Word documents with photos can be quite large. You can condense them to a PDF file, and no one can alter your PDF files if they fall into the wrong hands. Many agent websites will not let you upload documents of any real size, even though the trend in chef documents has gone heavier on graphics. Many of these agencies are still running websites that were designed in the Stone Age, and it can be quite frustrating to show a future employer how beautiful your work really is due to this.

Okay, you are legal to do this; you have no travel or driving restrictions, you are healthy, and you have completed your safety training. We now know what type of food you are most comfortable with and what your presentation is like; you now are ahead of the game. No one will take you seriously if you have not gotten this much done. Many would-be crew are not so well informed and roam the streets of private service capitals frustrated wondering why they are not getting anywhere. The industry is now too large for agents to take the time to help each would-be candidate, and to date there have not been many guides written about this ever evolving industry. Newbies usually meet working crew along the way and go off information based on that person's experience and trajectory; sometimes it's helpful, and sometimes it's not. If you have done this much you are already ahead of many other would-be crew.

Now it is time to draft up the most important document in your career: the venerable *maritime CV* (resume). Nothing is more important in your journey to employment than this little power tool. A marine CV is different from a normal resume in several ways. First of all, it requires a picture; this would be illegal on a land-based corporate resume. Second is that it will be in color. Third is that it includes less information about location and more info about you personally. When you are drafting a marine CV, you need to understand two things: What is the only purpose of this document? Who will actually be reading it and making the decision of hiring you? A successful CV is not designed to get you the job; its only purpose is to get you a face-to-face or phone interview— period. Your CV is going to fight a paper battle against other CVs. It needs to get the attention of the reader, present you in a clear and concise manner, and convey that you are worth contacting without the document self-destructing halfway down the page.

Let me give you the possible paths that your CV will travel on its way to getting you contacted. The first path would be through a yacht crew

Chef

Your Name
001-954-555-8888
Chef@email.com
USA Passport
STCW-95
Non-Smoker

Objective
To obtain a Chef position on a well-run motoryacht that does some chartering.

Qualifications
*Graduate of some awesome food school *8 years hotel & restaurant experience *STCW-95 *lifelong sailor * lifeguard certified. *Useless Bachelors degree from a state college in an obscure field. *Visited 9 countries (list them)

Experience
March-May, 2014
M/Y Super Suarbear (142' Christensen), Chef. Bahamas: Private

January-March, 2014
M/Y Captain Fantastic (124' Delta), Chef. St Maarten/Caribbean: Private & Charter.

January-December, 2013
Some guy's name steak house, Racine WI, Executive Chef

January-December, 2012
Trendy named restaurant, Duluth MN, Sous Chef

Personal Interests
I enjoy reading, swimming, sailing, Ninja Judo and dining out

References available upon request

FIGURE 21.2 SAMPLE CV/RESUME.

agent since you will visit them all your first day in town. Each agent has a format that they prefer, and most will not agree on one format. An agent has to decide if you are worth investing their time in and putting their reputation on the line for. Good crew agencies do not charge a fee so if they start the interviewing process with you it is a good sign. The agent will either bark orders to go and change things to their way, or if they like you will spend some time going over it with you (also a good sign). You may end up with several variations of your CV to please a few different agents. Some agents used to be department heads on yachts and have hired people themselves, but many have never worked on a boat or hired a crew member in their life. If a good agency wants you to make some changes, do it. The second route would be that an ally in the business is going to pass your CV on to a boat that is looking for a chef. That ally will be putting their own reputation on the line, so it needs to be tight. The third scenario is that you are putting your CV in a binder at a business relating to yachting as a free service to its clients. This is a chance to see how your document stacks up against others and how many other people are chasing the same dream. Department heads can come in and browse these binders at no charge and then can decide if they want to contact you directly. This is one reason that you will not put any sensitive information on your marine CV. The other route would be that you are walking the docks of big marinas or at an industry event and you pass your CV directly on to a captain (straight to the top!). That's a lot of people to please, but the only one that really counts is the person doing the hiring.

The avenues that put a middleman between you and the decision-maker are like speed bumps along the route to employment. Do not get me wrong: agents and allies are extremely important in stacking the deck in your favor as well as saving you from yourself in some cases. However, if the decision-maker does not see your CV or get who you really are, you are not going to get the gig. I worked for 15 years in private service, most of it freelance (short-term contracts)—I know what it is like to have a CV that functions and fulfills its purpose. Some loved my CV and some hated it, but the fact was that I got hired a lot over the years. I have also, in my time, hired quite a few people from sous chefs to stewards and house managers. I have also volunteered my time to helping many people go from frustrated to employed in our business. Throughout the years, I have watched countless captains and department heads go through the process of finding good candidates. When you apply to a hotel or restaurant, there is a human resources department or a general manager that will sit down with a pot of coffee

and spend a whole morning going over lengthy, polite documents trying to make sense of the real merits of candidates before placing them
in "maybe" resume piles. A few weeks later, they will redress some of
them and pass them on to department heads. It is rarely ever that way
with a yacht. The more likely scenario is that the boat has just finished
a hellish month of back-to-back charters and needs fresh crew. The
boat will head into port for a few days of replenishment before starting off on the next leg of its adventure. A captain or a chief steward
will call an agency and ask for three to five candidates each to replace
a burned-out chef, two junior stewards who are now going to work
with new boyfriends on their boats, and a deckhand who has finally
gotten a first mate position—oh, and a freelance engineer to cover for
the current one who has to take a refresher course to keep his license
current. That is four new crew members who will need to have all
their flights arranged and flown out in three days. That's 12 to 20 CVs
requested from one agency. But captains rarely request them from only
one agency, so that can be more like 40 taking into account that some
will be the same candidates represented by two or more agencies. This
then opens a political can of worms if they hire an obviously good
choice and need to decide who gets the hefty commission. Some of
the crew will have friends' CVs that they will want to toss on the pile.
The crew bridge is where the ships business is most likely to be done
with 20 to 75 CVs piled up competing for attention. By the way, few
captains will agree on what kind of format they prefer: too flashy, not
flashy enough, too long, too short.

That's why I am going to give you a basis to start from that is clear,
concise, and easy to get your head around. Remember what I told you
about this being the third most image-driven business? ADHD galore
and based on time over money—here's where it all comes together. If
your CV makes it to the bridge of one of only a few thousand of these
ultimate pleasure craft, you have won a huge battle but not the war. You
are now at least on the radar (maybe even literally). The good news is
that now you have been seen by the decision-makers. Even if they do
not take you this round, if they like your CV they may keep it on file
or even pass it on to a friend who needs a chef now. Yacht crew tend
to change boats more often than staff of any land-based business that

I know. A year is like an eternity on one of these boats unless you are the captain or the engineer. If your CV is memorable enough for the right reasons, the department heads who have seen it in the past may remember it when it comes around on another vessel.

But you really want to be hired now. The first thing department heads do when we print out the 75 CVs is to try to figure out what position each is for. There are men and woman in every position on these boats, so just because the picture is of a nice lady does not mean that she is not an engineer. We read enough of each to figure it out and then take a permanent marker and put the position in big block letters at the top right before making piles. It is important to have your name and the page number on each page of your CV so that when we print them out we do not mix some of your pages with those from other CVs. I like to save them trouble in the sorting by putting the word *CHEF* in big letters at the top right of the document.

The most important part of your CV is the picture: no picture = trash can; bad picture = trash can. It is the first thing we look at. If we do not like the way you look for any reason, we do not read any further. This is a very discriminating business. For this reason you may want to invest in having a professional photo done. You should be dressed in your chef jacket and shot from the waist up with you prominently in the foreground. It should be in color, and you should look like someone you would like to work with. This photo will go in the upper left and be of a decent size but not too overpowering on the document. We put this on the top left because it is human nature for people to read from left to right and we want the reader to see the image first.

For the lucky CVs that make it past the image round, we then look to the top right under the big word CHEF for your name and contact information and to see if we can legally hire you. Your contact info is your cell phone number and your email. Do not include your home address or current location. Anything you put on your CV is something that can be used against you. If we are receiving your CV in St. Martin and you are in Fort Lauderdale, we will fly you in. If we read a Washington state address, we may think you are there and pass on you. We will get a mailing address from you when you join the vessel and the agents will keep it on file. Your current location will change often; you do not need to put it on your CV and then have to change out your updated CV with each agency around the world every few weeks or months. Worse would be that we think you are not where you actually are and have you miss out on opportunities.

By being legally hirable, I mean three things: Can we take crew from your country of origin? Are you legal to charter? Do you have the necessary visas to work with us? You will have to include your country of origin; this is the country where your passport is issued. If you have dual citizenship, put both countries here. Next you will put STCW-95 here so we know that you are legal to charter. Some insurance policies do not allow their vessels to hire crew without one. Visa requirements vary, so do your homework before wasting anyone's time. You will put your current visa here: just the type of visa, not the number. Do not put your passport number or any other information that could result in your identity being stolen. This CV is an advertisement for you, not your whole life story. If you do not smoke or have no visible tattoos, you can put that here too; those are big plusses; however, do not if you smoke or are all inked up. All this information goes to the right of your photo like bullet points, larger than the regular text and in bold.

Okay, you look the part, you are clearly a chef, we know your name and how to get a hold of you, and we can legally hire you. Congratulations, you have made it through round two. By this point, the trash bin is getting fuller from less organized candidates, so good job. Next should be a titled section called *OBJECTIVE*. Make the title a little bigger than the regular text here, too and make the title bold. Your objective is a sentence that tells the reader what you are looking to accomplish and where you want to be in our industry right now. This is where we see if what we are offering is what you are looking for. Choose your words carefully. You have to understand that if you made it past the first two rounds the reader is now rooting for you. The reader also represents the boss and has an obligation to throw the CV in the trash as soon as the document does not corroborate its own story or has red flags or even if it is an excellent document that does not match up with our current needs. If you put that you want to be on a motor yacht that does charters but you do not have your STCW-95, you cannot legally do charters so your CV contradicts itself and goes immediately in the trash. If you want to be the chef on a formal motor yacht but in your photo you are in a baseball cap, untucked polo, and shorts with no food in sight—hello trash can. You should be as direct as possible. By this time you should have talked to other people in the business and figured out if you are looking to be a serious chef on motor yachts or a just have fun as a sailing chef. Also, you should have made a decision as to if you want to work with just one family or get onto a boat that takes on the challenges of chartering. Do not make the amateur mistake of saying, "I want to reach my full potential by pairing my love for people with my love of travel while offering my best, blah blah blah." That just says no experience, no direction. Be direct: "I am looking for a chef position on a motor yacht that does some chartering and travels extensively." Or, "I am looking for a head chef position on an estate that has a busy program and does a lot of entertaining." Avoid pigeonholing yourself, as in, "I am looking only for a chef position on a St. Martin–based boat that does not leave the dock much during the season." Make your desires known, but leave yourself enough room to be hired by as many programs as possible without compromising your career.

Keep the word count down, and get to the meat of the matter. Your first marine CV should be one page or two if you have a little time in your field. Yachties get bored easily, especially when hiring crew is one of many things on our to-do list that day. As world travelers, we get our heads around things fast. It should not take us five pages to get to know what you are about, especially if you are new.

The next stage is important and may be a little different from other CVs. The next bold headlined section is *QUALIFICATIONS*. This is your chance to say what makes you uniquely qualified to pull off your stated objective. This is usually short and sweet, with bullet points. Think of this as a 30-second reply to someone who is interviewing you asking, "What is your background?" You may say, "I graduated from so-and-so culinary school, I have been an executive chef for the last five years, and I have my STCW-95 [yes repeat this here], I grew up on a lake, and did lots of boating growing up. I have traveled all over Europe and Asia. I won a culinary competition in France and I speak three languages." Whatever pertains to the position and circumstances. Your travel experience is important. If you have no yachting experience yet, your small boat experience counts. Later in your career, it will not. We frequently have to work in environments that operate only in foreign languages, and a little experience in that helps. Put your other educational experience here, too. We love educated people, even if it is in other fields. It shows that someone somewhere has tested you and you had the discipline to challenge yourself even when not being paid.

The CV is like the layers of an onion being peeled back: this is what I look like; this is who I am; this why I am here; this is why I am uniquely qualified; and now these are the details of how I have spent my working life. The next bold titled section is *EXPERIENCE*. You will start at the most recent experience and go backward. Each entry will include the dates, the name and location of the business, your title, and any nonobvious duties. If you were the grill cook, you do not have to put a list of the meats that you cooked or that you set up your station and cleaned it afterward. You do not become a department head by being obtuse. Save those details for the interview if asked about them. The most important entries will be any boats that you have worked on, even if was just day work and not as the chef. We mostly care only about our own industry because it is so unique. We like to know what boats and crew you have a history with. When you enter a boat on your CV, put the dates, the type and name of the vessel, the length and make of the boat, then your position and the geographical locations traveled with that program:

> May–June 2014: M/Y *Lady Sugarmama* (174' Feadship),
> Chef, Bahamas/Virgin Islands.

The M/Y stands for motor yacht; S/Y would be sailing yacht. Most captains look directly at the length of the boat first as that determines whether or not you have experience on that class of boat or not. Next is the manufacturer of the yacht. A Feadship or Lursen yacht can cost three to five times the value of another yacht of the same length, will demand a more experienced crew, and will run with higher expectations. Unless you are very young, you do not have to enter jobs from college or before, and you do not have to enter secondary part-time jobs like dog walker or mall Santa. (I take that back—mall Santa is cool.) You may have had some downtime over the years; you do not have to highlight that. This is not a police statement, just an advertisement to whet the appetite of the reader to want to reach out to you. You can explain those times in person or over the phone. That being said, do not put anything on your CV that is not true; in our industry, we need to know we can trust you.

Once you have nice, clear, easy-to-understand, and concise entries for your work experience, we can move onto our last bold, titled heading: *PERSONAL INTERESTS*.

This section is also a little unique to private service because we not only work but also live together. It is important for a captain, who is trying to build a good crew, to have people onboard whose lifestyles complement each other. What we are looking for is *five* things that you like to do in your free time. Do not put things that are impossible in yachting; we want to know how you are going to spend your downtime. If you like reading, visiting art galleries, hiking, swimming, and tennis, cool. Maybe you will fit in with a healthy living crew. If you are into darts, billiards, and all-night clubbing, maybe you fit in with a harder partying crew. They both exist. It is important to give the captain a fighting chance to build a good team. Say what you really enjoy doing now, but keep it to five things that are not too weird.

The last line on your CV should say *References available upon request.* A separate document should have the names and contact information for at least five good references. As time goes on you will drop older, land-based references and add the yachting ones exclusively. Department heads do not like to talk to regular townsfolk and have to explain who we are and what we do. We may do this for a good newbie, but it's how you perform on the boat that is important.

There you have it: This is what I look like; this is who I am; this is why I am here; this is why I am uniquely qualified to accomplish that; here is the experience to back it up; here is what I do with my precious free time; and here are others who have witnessed it. If you get it right and you get this into the right hands, you now have a real chance at getting to the interview stage.

Before we discuss this, you should know that all yacht owners also have homes and you can certainly get estate jobs through yacht agents or from connections within yachting. Yachting is far more demanding than estate work. Many want a chef with yacht experience and seek out the few of us with it. Your marine CV will work for estate work; however, a simpler version in the same format will work for estates if a life at sea is not for you. You do not need an STCW-95 or any references to personal boating. Estate agencies work much like yachting agencies. One difference when hiring for estates is that people tend to take more time in the hiring process. Yachts do not usually make the call for crew until it is go time. Estates may put you through weeks of phone calls before actually getting a face-to-face meeting. Estate agencies come in two flavors: local and global. Southern California has the most number of private chef jobs. If you want a job there, go there. If you want one in New York City, go there. Without meeting you, nobody is going to hire you across the country when there are qualified local chefs who already live there and know the lay of the land, banging on their doors. If you are a rock star, Michelin star chef with impeccable references, it is possible, but the agencies still want to meet you in person before they put their credibility on the line. Global estate agencies do exist for the big players; they are based in places like Beverly Hills and London but still prefer to find candidates already in the regions where the jobs are. Estate jobs are more livable and do not turn over as much. There are far more estate jobs in the world than there are yacht jobs. They do tend to pay a little less, and having taxes taken out of your check is almost a given. Nonetheless, from this point on the interview process is the same, with the caveat that when you get past the phone call stage you may often be meeting with the boss himself. When interviewing for a yacht job you will almost always meet with the captain. Occasionally, a chief steward or the outgoing chef may talk with you first.

The first thing that you need to understand is that if you are getting a face-to-face interview with the decision-maker your odds are most often one in three or better that you have the gig. No matter how many unemployed chefs are roaming the same streets as you or how few jobs are available, you have to keep in mind that you do not need all available jobs—just one. When you get a fish on the line, and I am talking about a face-to-face, drop everything and pay attention. That interview should not be taken lightly. Even if you have multiple interviews that day or week, treat each one with all the focus and respect you can muster. I have seen people with a handful of aces lose the pot and watch all those gorgeous opportunities cruise out of the harbor without them simply because they did not focus.

You need to be ready for a few things when going through the interview process. One, you will not want some of the jobs you will be interviewing for, or they may not be your first choice. Second, at entry level, the person interviewing you may not be well organized, may not be very good with people, or might be downright disrespectful. Be graceful, maintain your dignity, and never be combative. Third, be prepared that at the entry level you may not be interviewing for the position in a private, dignified setting. I cannot tell you how many laid-back or uneducated captains will still interview you for a very high-paying or high-profile job in a bar or bagel shop. They will talk loudly in front of everybody about your experience, salary, and personal habits, often in front of people who you may know. It can be very embarrassing or off-putting. If you are a female, many smaller boat captains may want to hire you as part-time professional/part-time playmate. Be ready for this; it will inevitably happen at some point. Being a captain can be lonely, and when you have a small crew or it is just the two of you, not socializing as you travel is not an option. You will eat together, work together, and on some small boats even share a cabin. If it is someone that is fun and clearly just likes your energy, this can be a great way to get started in the business. If you get the vibe that a creepy captain is trying to put you in an uncomfortable or controlling situation, trust your instincts.

No matter how or where you get started in the industry, you are going to have to start somewhere, and you want the largest number of options possible. Each interview that you go through, you should come out with the best options for yourself without having burned any bridges. Consider the hour leading up to the actual face-to-face meeting as the starting point of the interview. This time should not be shared with any other obligation. Shower, dress appropriately, and start to think about the details that have been provided to you. Put the interviewers name in the foreground of your memory. Think about the job itself and what questions you may have for them. Keep in mind that your goal is to put your best foot forward and come out looking good. Even if you are not the right candidate for the gig right now, you can keep the door open for the future; any reports that get back to the crew agent, if one was involved, will be positive. Remember that even if they offer the job to you at the end of the interview, you may ask for a few hours or even in some cases a day to let them know if you are going to accept the position. Saying that you are interviewing for other jobs would be like picking up a girl for a first date and telling her about the other girls that you are seeing while you are on the way to dinner. Most people will understand if you have another interview scheduled for later that day or the next morning. Most, however, will not wait five days for you to interview with someone else as these offers are often time-sensitive and other candidates may not still be available after a few days and the whole hiring process will have to be restarted. Having too many opportunities can be just as stressful as not enough. Keep in mind that you can refuse an offer for employment; go for the brass ring and do your best.

Proper dress for an interview depends on the program itself and where the interview is. A tie is never worn in the yachting industry at any level. If you are going to be interviewed on the vessel itself in the Caribbean, on a 95° South Florida day or in a shipyard, a clean, ironed polo shirt and khaki pants or shorts will do. If the boat is fancy or you are interviewing in a boardroom at one of the agencies, maybe a nice, starched, button-down shirt, slacks, and a fine pair of loafers would be more appropriate. Never wear to a boat shoes that have to be tied and untied, as you will have to remove them. You cannot underestimate how awkward it is to have someone try to do this on a hot, sunny day with their sunglasses sliding down their sweaty nose on a bouncy, floating dock while trying to keep their balance—not a good first or last impression. When the boat is not in the area, sometimes the captain will fly into a yachting capital to interview crew in person and will be staying in a nice hotel. In this case, if you are meeting the captain in the lobby or a bar of a fine hotel, a blazer is not going to be out of line—but never a tie. If you are a lady chef, dress nicely (business casual) but not like you are going on a date, or you may end up on one instead of a job interview. If the captain offers a drink, choose club soda. If the captain orders a cocktail first and insists, have one—ONE. And leave a little in the glass when you leave. I know a big boat captain and his wife who are work hard, play hard types and drink most days. He will interview crew only out on the town to see if they can keep up. It is not ideal, but if you are going to travel with them it's best for both of you to know up front.

It's a strange business. The first time you walk onto one of these float-ing superpalaces, you are not going to believe what you are seeing. The interiors can be worth tens of millions of dollars, and crew will be scurrying everywhere. It may be a crazy scene. ADHD types are drawn to and do so well with this; there is so much stimulation every-where you look that it commands a level of superfocus to get anything done. Things get done faster when superfocus is applied. In short, your interview may be very different from any other that you have had. Concentrate on the person and the conversation at hand.

When interviewing for an estate job, you may be interviewed at the property itself, at a hotel, or at one of the boss's other assets, which may be a business or even a jet while it sits on the tarmac. I have seen it all. The first time you go to a penthouse in Manhattan, through the gates of a Palm Beach mansion, or up the hill in Montecito can be intimidating. Meeting a billionaire on their home turf is not something most people have experienced. Remember why you are there, and keep your shoes on for this one unless otherwise instructed. No shorts or khakis for this scenario, and you can have laces on your shoes.

Yacht or estate, bring a notebook and a pen; do NOT bring a copy of your CV. Surprised? If you hand them a CV in the beginning, it is considered an insult that you think that they do not already have it or that they have not done their due diligence. If they do not have one, you are at an advantage. Remember that your CV exists only to get you the face to face; its purpose has been fulfilled—you are there! At this point, anything on that CV can and will be used against you! You came for a face-to-face, not a face-to-the-back-of-a-piece-of-paper. Now is the time, here is the place: no piece of paper can explain you better than you can explain you. Advantage interviewee. Read the interviewer, and answer the questions conservatively. What is the worst thing that can happen? You tell the truth and are not the right candidate? You do not want a job that is not a good fit anyway unless you need a break to get into the business or are desperate for cash. If they offer a bottle of water or a cold beverage, accept. When meeting the captain or the boss, say hello and follow their lead; these types like to be in charge. If they offer a hand to shake, do so; if not, a simple how-do-you-do will suffice. Sit comfortably but not relaxed, make eye contact, and let them do the talking. You are to speak when spoken to, and your answers should be concise and professional. This is not the time to try out your new comedy routine or make polarizing statements. The big picture here is that they have already done a lot of legwork to get you to this point, and they are rooting for you. At this point, the job is yours to lose. Personality and chemistry count here: one you can control, the other you cannot. If you are polite, prepared, and professional, you have done all that you can do. If the boss starts telling the details of the job, take one or two word notes and at the end do not be afraid to ask intelligent questions. Avoid personal questions about the family or their fortune; you may ask a captain what field the family is in but not the family name or the name of their company. Do not ask the boss about their business. They will usually mention it if they feel comfortable. Do not underestimate how nervous you may be in this strange new environment; there was nothing at culinary school or in my hometown that would have prepared me for this. You will get more used to it over time. When the interview is over, either they will make you an offer, or they will thank you for coming and tell you the outcome at a later time. Even if you cannot wait to get out of there and are so glad that you will not be joining the program, keep your act together until you are fully out of the area. Do not speak of this to anyone but your closest confidants. If an agent was involved, you will want to find out how it went from the interviewer's side. It is very important to know how you come across to others in the business so that you can hone your interviewing skills and public image. If the agent asks you how it went, tell the truth but do not get overly emotional about the experience. No one likes drama in this business. Remember, it is you and the agent against the world, not you against the agent and the world. That being said, the agent needs to know that you are going to be successful at this and that they are not wasting their time on someone who is never going to make it in this field.

Okay, I have covered the in-person interview process as well as I can for someone new to this. Take each interview seriously, show respect, and take advantage of the face-to-face time as this may be the most time that you ever get one on one with the captain or the boss.

As I have stated, things can happen fast in private service, and crew changes are rarely ideal. A large percentage of the time, crew are changed on short notice when the boat is on location or the estate is in full service. You are not going to get a face-to-face interview. Many times the phone will ring and you will have a frantic captain or estate manager in desperate need of a chef RIGHT NOW! This is one of the hardest kinds of interviews. These calls are usually from some other part of the world by satellite phone with a slight delay. The captain will almost always have a thick accent when being phone interviewed, and the overall effect can make Charlie Brown's teacher sound like a diction coach. The hardest part of being a yacht chef is that you must always be ready; you are on call 24/7, 365 days a year. If a billionaire's chef gets in a car accident at 3:00 a.m. coming home from the club while on charter (yes, that can happen), someone has to get in there within three meals. Your bags are never fully unpacked, you do not buy concert tickets until the night of the show, and if your phone starts vibrating with an international number flashing while you are having free lunch and a few beers at a loud, South Florida strip club you had better have chosen a seat where you can get out the front door, get your sunglasses on (I hear those places are kept pretty dark in contrast to the South Florida midday sun), and lose your buzz within three rings. If you do not pick up the first time, the captain will have a list of candidates they will call in rapid succession. In a pinch, the first one to pick up gets the gig. That can be tens of thousands of dollars you missed for a free lunch. If you can get your head around what the broken-up caller is saying, you will be on a plane in 1–48 hours to a job that could last for months. Your storage must be organized, and your clothes and uniforms had better be out of the cleaners. For over a decade, every time the phone rang the hair on the back of my neck would stand up and I got tense.

Another awkward kind of interview is the non-emergency scheduled phone interview. The captain or the boss will be calling you at 3:00 p.m.—or, even worse, between 3:00 and 5:00 p.m. (a range). Make sure that you are in a place that you will not be interrupted. Sitting and staring at the phone as the clock ticks past the scheduled time can be really annoying. Treat the phone interview like a face-to-face, but instead of eye contact you must really listen to the tone and inflection in their voice. Keep in mind that if you have made it to this stage they are already interested; you do not have to oversell yourself. Your CV is your marketing tool; just live up to it without selling when you are already sold. Phone interviews are tough, but in an international field you will do plenty of them.

A last kind of interview you will face on occasion is rarely practical but can be fun: the cooking interview. Some owners and captains swear by them, and they are a good idea for someone new to the business. You will be asked to come either to someone's home or to a boat and do a trial dinner for two to eight people. This is so they can see how you organize, shop, cook, clean, and handle pressure. They will give you a certain amount of cash or a credit card; you will plan a menu—usually three courses—and make it happen. Often you will be asked to serve and introduce each dish. Normally, you will be paid a modest day rate for this, but if you get the job and are starting soon after they may just add an extra day onto your paycheck. It is important to realize that this is not a competition; pulling off a delicious, but not too over-the-top, menu with grace will be more important than pushing yourself to the outer limits of your culinary talent. When the meal is over and they ask you to sit at the table to have a glass of wine with them, do so. Keep in mind that once you are deemed competent the face-to-face is the most important part of the process. Trying to impress them with your commitment to detailing and seeing your guests only as they are leaving is not ideal. Plus, you should work clean and clean as you go anyway. Surely leave no trace that you have been there, but someone else will detail after you leave. Wear your full, clean chef uniform and be personable.

Wow, you now have your act together, your paperwork is immaculate, all your safety training certificates are complete, you have an idea about what agents do, and you have chosen what route and city you want to work in. You will now plan your attack and buy your plane ticket or pack up the car. Before you leave, you have one more step to go through. You are going to have to register with all the agencies in your field before you leave. You can look online at which agencies are active in the area that you want to work in and fill out the online applications. I'm not going to lie: this part sucks, and most of these websites are technically out of date and a test of patience and hair retention. There are trick questions on these, and no two are alike. Remember, anything that you put down here can and will be used against you, so avoid most optional fields. You do not have to put your measurements and whether you smoke unless you are a gorgeous human being and a nonsmoker and want to brag. Do not put your passport number in—just the country of origin. You can give these important details at the face-to-face with the agent. Some scams have been aimed at our industry, and these websites are far from totally secure. Keep your social security number to yourself until you are being hired. Some of these websites may be down or broken, and some will not allow you to upload anything over a few dozen kilobytes. Muscle through and feel free to call them if you are having a major problem. Sometimes the secretary will have you email them a file bypassing the site and they can add it on their end. Most agencies will not see a newbie unless you have all your paperwork in order and have fully completed the online application. Next you will want to schedule in-person interviews with all the agencies in your planned area of attack. Estate markets are not organized and do not follow seasons as much, so if you want to work in Los Angeles just move there, rent a room somewhere, and throw the spaghetti at the wall. Be persistent with the agencies to let them know that you are hungry. Keep in mind that these people in no way owe you a living and that bringing them homemade cookies or lunch for the office can go a long way.

In the yachting world, you have to want this badly. You could go to Newport, Rhode Island for the summer; however, it is a short, albeit superfun, season. If you are rich or already in Europe for the summer season, Palma, Mallorca and Antibes, France will do just fine for job hunting. St. Martin is the undisputed winter season yachting capital, but for the most international bang for your buck year round there is no substitute for the yachting capital of our solar system: Fort Lauderdale, Florida. This city features many crew houses and small businesses specializing in crew housing, which charge a reasonable fee for a bed in a shared room with other yacht job seekers and crew that are taking short maritime courses, such as an STCW-95. The feel of most South Florida crew houses is reminiscent of any episode of MTV's *The Real World*. You may choose to arrive in Fort Lauderdale, get booked into an STCW-95 class, and arrange for interviews with the yacht agencies that begin as soon as you have completed the course. You can book into a crew house for the whole process.

When interviewing with the agents, treat it just like you would a job interview. You can be a little more personable, but they like people who are organized, fun to be around, and committed to the cause. Avoid anyone outside of the yachting business during this period, as any energy focused outside of the business will make you less competitive. Go where the yachties go, as this is where you will make connections. Do not date anyone during this process because that is not why you have spent all this money and traveled this far. Yachting capitals are sunny, sexy places with lots of distractions. Keep focused on why you are there.

Once the agents know and like you, go in person to check in twice a week. If you are staying around the 17th Street causeway, you will be able to walk to most. Attend all yachting events and read the crew and trade magazines. Live and breathe this life. Be good to your fellow yachties and dress like they do. Only wear flip-flops at the beach. Do not become known as a drunkard: we all have big nights from time to time, but Monday through Friday from 9 a.m. to 5 p.m. stay sober.

Once you have this much down, you have the odds stacked in your favor. The agents are one of the only parts of the hiring process that you have some level of control over. Social media and word of mouth have overtaken the agent game, and more jobs now come through social media and who you know than through agencies. There are groups on social media websites like Facebook for yacht chefs, estate chefs, and all forms of private chefs. People will announce that they are looking for a chef for a 40-meter motor yacht based out of Monaco and then give you an email address to send your CV. This way no one has to pay a hefty agent fee. Some captains will always use the agents because this is a cover-your-ass business and if the chef does work out the captain can always say that the chef came highly recommended by an agency that knows them, has checked their references, and keeps a file on their career progression. If a chef does not work out and does not last 90 days, most agencies will replace the chef with a new candidate at no charge. Looking for a chef can be a lot of work while you are at sea or in guest service. The agents are not going away.

Now it is time to take your destiny into your own hands. You are going to go to networking events, walk the docks, and tap on the hulls directly, and you are not going to be walking around with a stack of CVs. You will send your CV as an attachment in an email to yourself titled Chef (your name's) CV. Anytime that you meet someone who is a captain or an agent, who knows someone who is looking for a chef, or who is even just a more experienced yacht chef than you, you can forward the email from your smartphone right on the spot, no matter where you are in the world. You can also print it off that email.

The real marketing tool you have all the time is going to be your business card. You must have plenty on you at all times! They must be in the same colors as your CV, and they must have the same picture as your CV. I got this idea from real estate agents. They realized that as a real estate agent what they were really selling was themselves: Who do you want to go through this process with? The successful ones always had a picture of themselves, not a house that took up half the card on the left side, for reasons mentioned earlier. On the right hand side of the card I put the word CHEF in big letters, my name under that, then my cell phone number and my email under that. Finally I put in big letters *YACHTS & ESTATES*. That's all anyone needs to know. Make sure that they can read everything easily and that the whole card pops! This is not a mini CV; it is an appetizer that makes you hungry for the CV.

FIGURE 21.3 SAMPLE BUSINESS CARD.

You see, at most yachting events there are tons of people and the booze is free and endless. To deal with the crowds and all that energy, most people indulge in a few cocktails. I was a well-known person in our business, so at most events I would receive on average more than 20 business cards per night. The next morning, I would sit at my desk and go over the stack to see who I had met and recall the conversations we had. I would seek out the good ones on Facebook to keep in touch as we traveled. The problem was that so many had no picture or it was too small. A new deckhand or steward would be great and I would love to help them, but their card was too conservative and featured a beach scene or a picture of a generic boat. I would have also met some terrible people that night, whose cards also featured an aquatic motif. I could not remember who was who, so good or bad: no picture = trash can. I simply could not take the risk. Your business card should be handed to anyone in the business who is a captain, department head, employed, or just plain cool. Do not offer one to creepers or douche bags. It is also used as an ID at yachting events, shipyards, and purveyors you are dealing with.

You now have way more information than my peers and I ever had when we started, and we made it happen. That's because we knew that yachting and private service in general must be entered into with reckless abandon or not at all. You should give it everything you have and leave nothing left for the swim home. You must use common sense and be able to handle rejection. You must be grateful when you have success and humble with your failures. You must treat your fellow chefs like fraternity brothers and sisters and be fearless in your pursuit of career satisfaction. It is an amazing feeling when you finally make it and everyone simply introduces you, "This is my friend, a chef to billionaires!"

INDEX